SpringerBriefs in Operations Research

SpringerBriefs present concise summaries of cutting-edge research and practical applications across a wide spectrum of fields. Featuring compact volumes of 50 to 125 pages, the series covers a range of content from professional to academic. Typical topics might include:

- A timely report of state-of-the art analytical techniques
- A bridge between new research results, as published in journal articles, and a contextual literature review
- A snapshot of a hot or emerging topic
- An in-depth case study or clinical example
- A presentation of core concepts that students must understand in order to make independent contributions

SpringerBriefs in Operations Research showcase emerging theory, empirical research, and practical application in the various areas of operations research, management science, and related fields, from a global author community. Briefs are characterized by fast, global electronic dissemination,

standard publishing contracts, standardized manuscript preparation and formatting guidelines, and expedited production schedules.

Eugene J. Zak

How to Solve Real-world Optimization Problems

From Theory to Practice

 Springer

Eugene J. Zak
Independent Consultant
Redmond, WA, USA

ISSN 2195-0482   ISSN 2195-0504   (electronic)
SpringerBriefs in Operations Research
ISBN 978-3-031-49837-4   ISBN 978-3-031-49838-1   (eBook)
https://doi.org/10.1007/978-3-031-49838-1

This Springer imprint is published by the registered company Springer Nature Switzerland AG
The registered company address is: Gewerbestrasse 11, 6330 Cham, Switzerland

Paper in this product is recyclable.

To my beloved wife Olga, and our cherished children, Konstantin and Katherine

Preface

Transitioning from theoretical research to the dynamic world of real-life practice involves navigating a multitude of intricacies, encompassing a comprehensive understanding of customer requirements, proficient modeling, the development of efficient algorithms, the application of specialized solvers, creation of commercial-grade software, as well as the ongoing aspects of maintenance and support.

The objective of this book is to serve as a guide for Operations Research practitioners with academic backgrounds who launch or advance their careers in industry or business. Based on extensive experience in large corporations and startups, the author has distilled the essential principles and insightful tips underlying successful practical projects. These principles are illustrated through real-world optimization problems drawn from original cases or the authors' earlier publications.

Redmond, WA, USA   Eugene J. Zak
September 2023

Introduction

*A lesson learned the hard way is a **lesson learned** for a lifetime.*

Working in the industry on both sides of the Atlantic, employed by high-tech companies, I have accumulated significant experience in solving practical Operations Research (OR) problems that, in most cases, are not textbook problems studied in universities and colleges. Still, academic textbook problems are often the starting points for all research and development performed in the industry. If an academic motto is "Publish or perish," an industry motto is "Monetize or necrotize." If we include non-profit organizations, the motto becomes "Bring value to the customers."

When you start writing a book, an obvious question arises: Who is the target audience? Who might be interested in reading this book? In my case, the answer is straightforward—my prime audience is OR practitioners who landed their career ambitions in the inspiring world of industry, business, and service. Saying that my prime audience is OR practitioners, I also keep in mind that the real-world formulations of the classical OR problems also present a considerable interest for the academic community and students.

To the best of my knowledge, this book is the first attempt to illuminate the similarities and differences in academia and industry R & D on the level of mathematical modeling of real-world problems. It is hard to publish your results when you work for a company. First, you may not have time to prepare a paper; second, management may not consider this time productive. And third, the most important one, companies guard their intellectual property closely—your work belongs to the company. In a few cases, the companies allowed me to publish and make presentations at scientific conferences, either for publicity or to keep the current staff and attract new employees. I see another positive side of publications even when you work in the industry—the feedback from the broader research community, which may even improve your program product.

My first assertion that I will push through the whole book—you cannot be a successful OR practitioner if you do not have a solid theoretical background. Being a good theoretician is a necessary but insufficient condition (as we sometimes say in mathematics) to be a successful practitioner. Ideally, you should possess a Ph.D. in Operations Research or Applied Mathematics. However, I met very talented practitioners with only a B.S. in Mathematics. Proficiency in mathematics and System Analysis is the cornerstone of your success as a practitioner.

You may ask—what is a sufficient condition to be a successful OR practitioner? In my view, it is an ability to get expertise in the subject area you want to work in. The Power System as a OR subject area is complex and takes significant learning time. I was glad to have an Electrical Engineering background when I started working on modeling in the power industry. If you communicate fluently in the professional language of your customer, there is a high chance you will succeed.

When we talk about "industry," we should remember that every company has its specifics and its own "culture." The significant differentiating factors are the company size, age, ownership (public or private), and the type of project you will be working on.

Some projects are "internal" projects that serve the company's needs. For example, if you work on marketing campaign optimization, your customer might be a marketing department of the same company you are in. Practically, there is no competition in this case. It might be one—if the solution already exists and you are working on the next generation of the same system. Only large companies can afford this type of project.

Other projects—"external" projects—serve your customers that are outside of your organization. For example, your company delivers a crew scheduling solution to an airline company. It automatically implies you will compete with other vendors, and winning such a competition is challenging. External projects are the only projects in small companies. When I worked on MajiqTrim™ (MAJIQTRIM: Trim optimization software 1995) in the paper industry, our team competed with five vendors, and our customers selected the best solution for their needs. Based on my working experience in both types of projects, I conclude that "external" projects are more challenging than "internal" ones because of fierce competition.

When I intended to write this book, I asked myself if someone had published something similar. To the best of my knowledge, I am not aware of it. Even so, as there are no two humans beings with identical DNA, there are no two scientists with the same career path. I am sure my career path, which spanned academia and industry, over two different economic systems, over small and big companies in the machine-building industry, paper industry, technology companies, power industry, retail e-commerce, and cloud-computing industry, is quite unique. As a result, I assume I can provide insights that readers find interesting and valuable.

And one more note. I use the term "project" and not "case study," though there are a lot of commonalities between these two. The first is widely used in industry, while the second is common in academia. "Case study" does not necessarily assume that the result of your work is a program product; "project" does, or it aims at least for that.

I structured the book in two parts. The first part summarizes lessons I learned across all my projects. The second part illustrates those lessons using the most fascinating practical projects.

Reference

MAJIQTRIM: Trim optimization software (1995) Users' manual. Majiq Systems and Software, Redmond, WA

Acknowledgments

This book culminates three decades of dedicated work alongside professional teams at Majiq Inc., IBM, Alstom Grid (acquired by GE), and Amazon.com. I sincerely appreciate Chris Rennick, Michael Rainwater, Helmut Schreck, Russ Philbrick, Ricardo Rios-Zalapa, and Dale Henderson for their outstanding collaboration.

I am also grateful for the invaluable feedback I received on my talks and publications from esteemed colleagues within the Operations Research community: Zelda Zabinsky, Ed Klotz, Alexander Shapiro, John L. Nazareth, Eugene Levner, John Milne, Anatoly I. Karpov, Pavel Buzitsky, Anatoliy Rikun, Jose Goncalves, and Victor V. Shafranskiy. Additionally, the Operations Research forum on "StackExchange," particularly Paul Rubin, deserves recognition for their prompt and insightful responses to my queries.

I thank the Google OR-Tools, SCIP, and Gurobi teams for providing the essential modeling tools and MIP solvers necessary for solving optimization problems. I would like to express my gratitude to the INFORMS and Elsevier for granting me permission to use excerpts from my early publications.

Acknowledgments

Disclaimer

While I used ChatGPT 3.5 solely for language improvement, the book is entirely my creation from the first page to the last.

Contents

Abbreviations

AC	Alternative Current
AIMMS	Advanced Interactive Multidimensional Modeling System
AMPL	A Mathematical Programming Language
CSP	Cutting Stock Problem
CSV	Comma-Separated Values
DC	Direct Current
LCM	Least Common Multiple
LP	Linear Programming
MIP	Mixed Integer Programming
NLP	Nonlinear Programming
OR	Operations Research
SCIP	Solving Constraint Integer Program
SHP	Sequential Heuristic Procedure
SMMP	Slitter Moves Minimization Problem
SSP	Skiving Stock Problem
UCP	Unit Commitment Problem
WSSP	Warehouse Storage Space Problem

Chapter 1
Practical Tips

1.1 Master the Subject Area: Embrace Functional Requirements—Think Outside the Box

As I mentioned above, becoming an expert in the subject area is a preliminary condition for devising an adequate model of reality. The best way to understand the functional requirements is by putting them in writing and prototyping a computational framework; in other words, learn by doing. You can achieve it in any general-purpose programming language of your choice. I used C, C++, C#, and Python in most projects. The goal of this exercise is hands-on learning of the subject area.

If you are an expert in the subject area, you are a step closer to formulating a problem that will be a cornerstone of your project. Presumably, the following quotation belongs to American mathematician John Tukey: "It is better to solve the right problem approximately than to solve the wrong problem exactly."

An end-user knows what kind of output data he/she would like to get from your application. However, even a verbal formulation of the problem is your task. And you are equipped well for this task now. First, you are on par with your customer in understanding the solution space. And second, what differentiates you from your customer in a positive way is you are an expert in OR and System Analysis. It is essential to draw a boundary of your system and to understand its place in a broader picture of the universe.

A possible verbal problem formulation is: "Given the requirements A, B, and C, find the minimum of the functional D." This formulation leads to an optimization problem. I argue that most real-world problems are optimization problems. Even in an everyday situation, say cooking your breakfast, you solve an optimization problem: "Given the food available and the culinary restrictions, cook the most enjoyable/useful breakfast."

More formal optimization problems formulations are:

- Routing problem—"Given the transportation network and the current traffic conditions, what is the optimal route from point A to point B?"

E. J. Zak, *How to Solve Real-world Optimization Problems*, SpringerBriefs in Operations Research, https://doi.org/10.1007/978-3-031-49838-1_1

- Planning problem—"Given the manufacturing capacity and customer orders, what is the best plan for production?"
- Scheduling problem—"Given the set of machines and operations, what is an optimal way to assign operations to the machines?"

As an OR practitioner, you know well all "classical" optimization problems described in numerous textbooks. As Kurt Lewin, a pioneer in organizational psychology, mentioned: "Nothing is more practical than a good theory." However, classical models and algorithms rarely provide a ready-made solution to a practical problem. Both steps are essential: acquiring knowledge of the subject area and formulating a real-world problem.

If several customers originate the same problem, try to work with all of them: even a slightly different perspective enriches your understanding of the problem space. It may facilitate the development of a better and reasonably generic design.

1.2 Think Beyond Rigorous Optimization: Develop Heuristic Algorithms

Let me start this tip by characterizing "model" and "heuristic." A mathematical program or *model* is a problem formalization when you define a restricted solution space and designate a criterion (or criteria) for picking one or more solutions as optimal. So, the model is a kind of intermediate entity between the problem and solutions. Though a general-purpose programming language can formalize the model, special-purpose *modeling languages* do it more professionally. Some modeling languages originate from their specific optimization solvers (e.g., Mosel works well with Xpress (FICO Xpress Mosel 2017), Concert Technology employs CPLEX (IBM ILOG CPLEX Optimization Studio 2017)), but other languages are solver-agnostic (AIMMS Optimization Modeling 2023; AMPL Development 2022–2023; Google OR-Tools n.d.). A modeling language formalizes the presentation of the input data, variables, constraints, and criteria.

A modeling language passes the model to generic optimization algorithms: primal simplex, revised simplex, and interior-point algorithms for solving Linear Programming (LP) models; branch-and-bound and cutting-plane algorithms for solving Integer Programming (IP) and Mixed Integer Programming problems (MIP) (Wolsey 1998); various iterative algorithms for solving Non-Linear Problems (NLP) (Bazaraa et al. 1990; Chvatal 1983; StackExchange Network 2019; Williams 2013). So, a model is not an algorithm but a presentation of the problem in a modeling language so that general-purpose optimization algorithms can work with such a problem presentation.

When discussing *heuristics,* I talk about direct algorithms that translate the problem data into a solution. Every problem may have a specific algorithm suitable for solving this type of problem. Why do we need a heuristic when we can use a modeling approach? The answer is simple: performance, on the one hand, and difficulty of some functional requirements, on the other. Typically, a

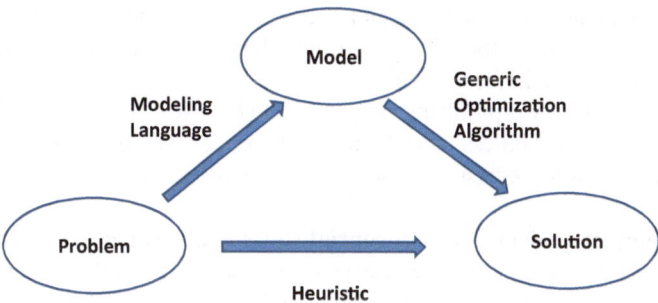

Fig. 1.1 Model and heuristic

custom-designed heuristic with a linear time complexity runs faster than a model with a generic optimization algorithm. The difference might be insignificant for small-scale problems but is critical for large-scale problems (Fig. 1.1).

Another vital aspect is early communication with a customer or an end-user. Getting proper guidance at the beginning is essential without wasting time exploring the wrong directions. That is why it is helpful to develop a reasonable heuristic approach as quickly as possible. Build your first system quickly and then iterate. "Quick-and-dirty" solution justifies your approach to the problem.

If a heuristic delivers a solution making sense to the customer, your learning process is on the right path. This first interaction with your customer often leads to further clarification of the functional requirements or even their revision. After that, the subsequent iterations follow. Ultimately, your understanding of the subject area and the related problem space is good enough for further work.

One of the approaches for developing a reasonable heuristic is to observe how this problem is solved manually.

1.3 Craft Intelligent Models for Real-World Problems

An excellent book by H. Paul Williams (2013) describes a model concept and recommendations for building efficient mathematical programming models. I will emphasize practical aspects of the effective or "intelligent" models. I define an *intelligent model* by a few inherent features that emphasize its creative aspects. Not in any order, these features are the following.

An intelligent model is as simple as possible. Following a compressed version of a famous A. Einstein's aphorism: "Everything should be made as simple as possible, but no simpler," we can state that an intelligent model is as simple as possible but not simpler. An iconic example of simplicity vs. complexity is the Copernican model of the Solar system vs. the Ptolemaic one. Positioning the Solar System center not on Earth but on Sun, as Copernic proposed, significantly simplified the planet movement equations. Closer to our topic example is the pair of the pattern-based model for CSP vs. the item-based bin-packing model. We will consider it in Chap. 2 in more detail.

Too often, real-world problems are complex. However, chasing modeling every aspect of reality is a fruitless task. An adequate model should include the essential elements and answer the basic questions. Maintain the balance between simple and complex.

An intelligent model has an appropriate level of abstraction (generalization). Avoid subject area specifics as much as possible. Keep in mind that the model is not rigid. It should be extensible to a certain degree. If a transportation problem concerns trucks, adding a new transportation mode, e.g., railroad, should not cause dramatic changes in the model. Again, keep the balance between abstract and concrete.

An intelligent model is suitable for several use cases. Try to expose problem specifics in data, not in the model. An appropriate degree of abstraction leads to tackling arbitrary levels of data granularity in terms of products, resources, and time. Particular interest may present a product aggregation, especially when millions measure a product's nomenclature.

An intelligent model is only sometimes apparent. For example, the bin-packing model is a natural candidate for minimizing the number of bins for packing items of different weights or volumes. However, this model could be more efficient. The explanation is that its LP relaxation is far from the convex hull. It took the ingenuity of P.C. Gilmore and R.E. Gomory (1961, 1963) to devise a pattern-based model and an elegant and efficient *dynamic column generation technique*.

Keep the balance between obvious and concealed.

An intelligent model is efficient and scalable. An intelligent model must run fast. Nowadays, performance is almost exclusively concerned with IP and MIP models. Even with a hundred of thousand variables, LP models are easy to solve, even for laptops. Granted, intelligent commercial and open-source solvers contribute to high performance. Still, creating integer models close to their convex hulls provides the foundation for their efficiency.

Ideally, the running time of a scalable model is a linear function of its input size. It might be the case with LP models. For example, a simplex algorithm on practical problems may take up to $3m$ iterations where m is the number of rows in the LP matrix (Nazareth 2004). However, the scalability of IP and MIP models is more challenging. Look for an MIP formulation with a tight LP relaxation. Data aggregation can improve the model's scalability.

An intelligent model is not hungry for the volume of input data. It means an intelligent model consumes a manageable volume of the input data to develop a decent solution. Contrarily, relatively compact input data is sufficient to deliver insightful results. Also, sometimes intensive preprocessing converts a vast volume of data to the manageable size required by the model.

An intelligent model is the core of all operations. Get rid of auxiliary operations not relevant to optimization. Move those operations into pre- and postprocessing. E. g., preprocessing can include input data normalization and validation.

Postprocessing can include converting the output solution to the original data space and calculating the solution evaluation metrics (solutions qualities).

1.4 Navigate Infeasibility: Blend "Hard" and "Soft" Constraint Formulations

In business applications, the approximate data accuracy is around 5%. The obvious implication is that we can violate a business constraint $\mathbf{a}^T\mathbf{x} = b$ to a certain degree. There are several ways to "soften" this hard constraint.

1. Adding a *constraint violation variable*:

$$\mathbf{a}^T\mathbf{x} = b + y$$

$$y \in R^1$$

Here, y is a constraint violation variable, and one of the objectives is its absolute value $|y|$ minimization.

In practice you can introduce additional non-negative variables $u, v \in R^1_+$ such that: $y = u - v$. Then minimizing $u + v$ leads to minimization of $|y|$.

2. Adding a *normalized constraint violation variable*:

$$\mathbf{a}^T\mathbf{x} = b(1 + y)$$

$$y \in R^1$$

Here, y is a normalized constraint violation variable, and one of the objectives is its absolute value $|y|$ minimization.

You can use the same linearization technique as in the previous case.

3. Adding a constraint violation variable either in the form of case 1 or 2, and using a convex function for the violation minimization.

Let us define a non-negative variable $z \in R^1_+$, and construct a piece-wise convex function of given points $(y_1, z_1), (y_2, z_2), \ldots, (y_n, z_n)$ and slopes s_1 and s_{n+1}. See an illustration in Fig 1.2.

To make an analytical description, we define segment slopes:

$$s_k = (z_k - z_{k-1})/(y_k - y_{k-1}), \quad k = 2, 3, \ldots, n.$$

The convex constraints:

Fig. 1.2 A piece-wise
convex function; $n = 5$

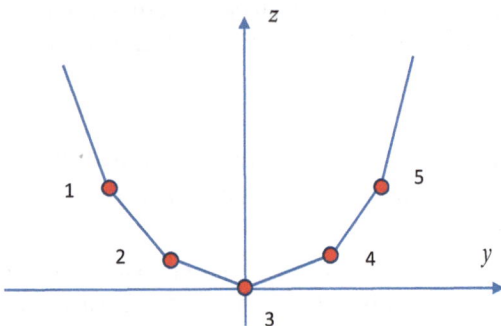

Fig. 1.3 A simple piece-
wise convex function; $n = 1$

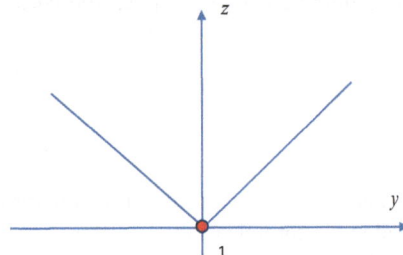

$$z \geq z_k + s_k(y - y_k), \quad k = 1, 2, \ldots, n$$
$$z \geq z_n + s_{n+1}(y - y_n),$$
$$z \geq 0.$$

The objective function:
Minimize z.
What is good here is a progressive penalty for the constraint violation.

By the way, we can present cases 1 and 2 also as a simple piece-wise convex function with a couple of constraints:

$$z \geq -y,$$

$$z \geq y.$$

See an illustration in Fig. 1.3.

"Soft" constraints allow us to avoid infeasibility. Getting an infeasible solution is a frustrating experience for an end-user. Conversely, a solution with non-zero values of constraint violation variables gives a user a clue about what is wrong with the input data or tight constraints.

1.5 Practice Multi-Criteria Models: Leverage Lexicographic Optimization

Criteria have two primary sources. First, they come from the problem nature, e.g., maximize customer satisfaction and minimize the cost of operations. Second, they come from soft formulations of constraints by minimizing constraint violation variables (see the previous tip 1.4).

There have been numerous methods for formulating and solving multi-criteria models. We can summarize them in two groups: folding all criteria into one and sequential optimization of one criterion at a time. From my experience, folding all criteria into one using the so-called "weights" is questionable. First, nobody knows what the weights should be. Moreover, certain weights may work well with one data instance and fail with another. In addition, the criteria nature is sometimes incompatible, so finding criteria commonality is impractical.

In practical applications, I often use *lexicographic optimization*. I can describe it as follows.

Let us specify n criteria: $f_1, f_2, \ldots, f_n$. Without loss of generality, we can assume that all criteria are minimization criteria. Also, the ranking (importance) criteria are in the same sequence. Then the lexicographic optimization is a sequential optimization of criteria $f_1, f_2, \ldots, f_n$ over the same model complemented by additional bounds on the optimal values of the previously optimized criteria (with an exception of the first one f_1). A flowchart in Fig. 1.4 visually presents the lexicographic optimization.

The advantage of lexicographic optimization is using natural optimization criteria and accumulating optimal solutions with every sequential run. Generating multiple optimal solutions is a valuable outcome of lexicographic optimization (see the next tip, 1.6).

The obvious disadvantages of lexicographic optimization are the following:

- You must rank the criteria,
- You may experience extended running time.

Still, my practice favors lexicographic optimization among other methods of multi-criteria optimization.

1.6 Expose Multiple Optimal and Near-Optimal Solutions

Usually, the LP/MIP solver returns an optimal solution though the potential number of optimal solutions might be infinite. Please refer to Fig. 1.5, where the minimized objective function position coincides with a face of the feasibility set (called the optimal face). Any point on the optimal face is an optimal solution.

There are special-purpose algorithms to explore all extreme points of the optimal face. However, for practical applications, I recommend introducing multiple criteria

Fig. 1.4 Lexicographic optimization

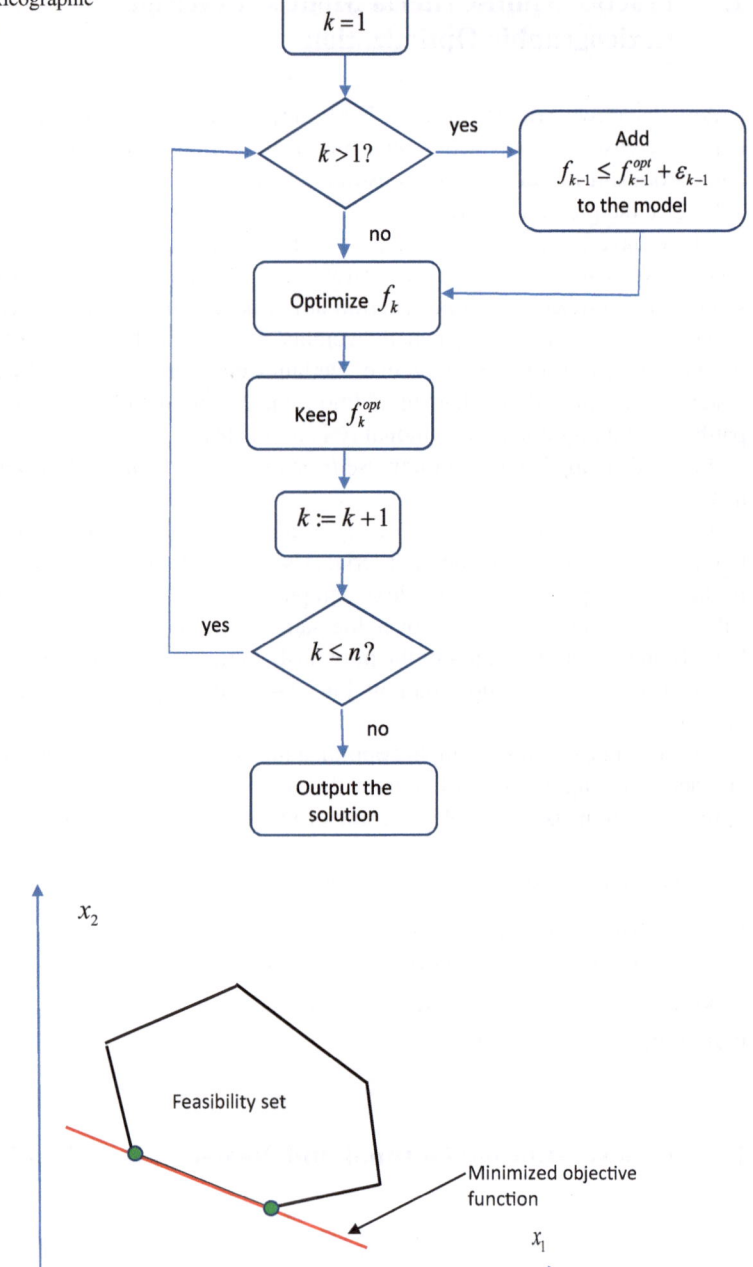

Fig. 1.5 Multiple optimal solutions in the two-dimensional space

with different rankings and applying lexicographic optimization (see the previous tip 1.5). Every pass of the lexicographic optimization enriches the pool of optimal solutions, valuable to the end user.

Varying some setting parameters, we can develop multiple near-optimal solutions. We can explore near-optimal solutions by running a heuristic. Moreover, employing various heuristics further diversifies a set of near-optimal solutions.

Why is it important to generate multiple optimal (and near-optimal) solutions? First, only some criteria can be effectively optimized. A good example is a CSP, where the prime criterion is the number of sets or waste. This criterion is a good candidate for the first-ranking criterion. However, another essential for an end-user criterion is the number of patterns. The existing CSP model cannot quickly minimize the number of patterns. That is why the pool of optimal solutions, especially those coming from the CSP heuristic, is essential to pick a solution with minimal patterns.

Second, since a model approximates reality, some secondary factors might be out of consideration. Given identical values of formalized criteria, two solutions may still differ in other secondary factors, which might play a role in the final selection by a customer.

Ideally, the system should offer several solutions to the customer with a designation of the "best" one.

1.7 Decompose a Problem into Manageable Interconnected Modules

The idea of a modular insight for large, complex systems is not novel. Programming paradigms such as object-oriented, functional, event-driven, and procedural programming widely use the modular approach. A modular approach applies to optimization problems as well (Conejo et al. 2006).

First, we should mention *generic decompositions* such as Benders decomposition (Benders 1962), Danzig-Wolfe decomposition (Bazaraa et al. 1990), Gilmore-Gomory dynamic column generation (Gilmore and Gomory 1961, 1963), and Lagrangian relaxation (Fisher 1985). For example, I used Benders decomposition to solve an industrial cooperation problem and other MIP problems. I applied the Gilmore-Gomory dynamic column generation for solving CSP.

The second type of decomposition is *model-specific decomposition*, which springs from the problem formulation itself. A good example is a slitter moves minimization while cutting big jumbo rolls into custom rolls. You cannot quickly solve the problem directly. An indirect approach is a problem decomposition into two subproblems:

- CSP, and
- Slitter moves minimization problem (SMMP) itself: Given the cutting patterns, find its sequence and slitter permutation for every pattern to minimize the number of slitter moves.

The second subproblem is a complicated combinatorial problem. We will consider it in Chap. 2.

Another excellent example of model-specific decomposition is the uncoupling of the production and transportation modules in the supply chain problem. The production module defines the specialization of every production unit. The transportation module sets logistical interconnections between the production units.

Contrary to the generic model decomposition, the model-specific decomposition does not always provide a global optimum. Having a single model that finds a global optimum is intriguing, but this is a trade-off between solution quality and computational efforts. Another factor you must consider is how easily the existing planning or scheduling practice will embrace a single or decomposed model.

1.8   Unveil Hidden Symmetry: Explore New Problem Counterparts

An exciting example of the complementarity of technological processes is a CSP. The underlying technological process is cutting wide jumbo rolls into smaller widths rolls. What is a technological process complementary to the cutting one? It is merging small rolls into larger ones by gluing or *skiving*. This process brings a new type of OR problem: *the Skiving Stock Problem (SSP)* (Johnson et al. 1997). A section in Chap. 2 is devoted to the SSP. This pair of "cutting–skiving" problems correspond to a similar pair of "packing–covering." Indeed, cutting and packing are essentially the same from the modeling perspective as skiving and covering (see Fig. 1.6).

Are there any other examples of such complementarity? It is an open question.

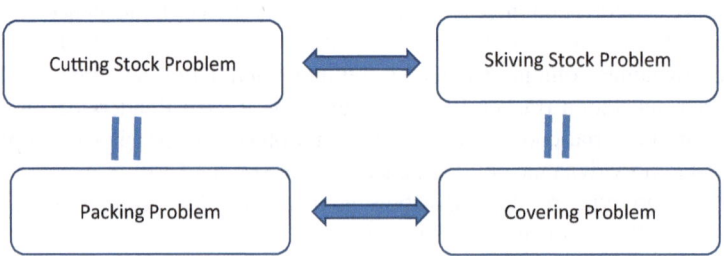

Fig. 1.6 Problems complementarity

1.9 Delve into Multi-Model Problem Resolution

Often a single model is sufficient for tackling the problem successfully. In some cases, two or more models must be in your technique's arsenal. The first reason is the existence of different use cases and different functional requirements. A model can be more effective for one use case and less effective for another one. The second reason, I mentioned earlier, is an accumulation of multiple optimal solutions.

Examples of two models solving the same problem are a pair of the bin-packing model and the pattern-based model for CSP (Fig. 1.7). Though the second one is much more efficient than the first one, small instances of the bin-packing model can be performed efficiently. We will consider both models in Chap. 2.

Another example is a pair of continuous and discrete models for the warehouse storage space problem (Fig. 1.8) which we will consider in Chap. 2.

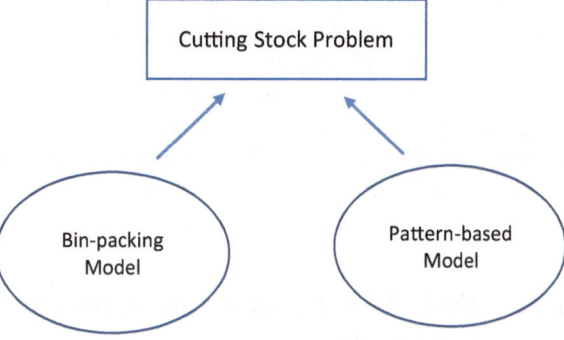

Fig. 1.7 Multiple models for the Cutting Stock Problem

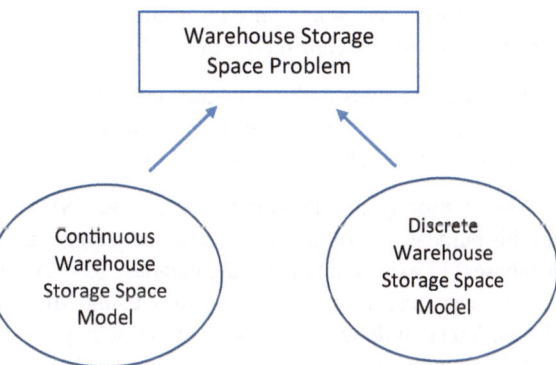

Fig. 1.8 Multiple models for the Warehouse Storage Space Problem

1.10   Deliberate Choice Deterministic vs. Stochastic Formulation

The question is: What type of formulation should you pursue: deterministic or stochastic? There is no definitive answer. Everything depends on the problem's nature and the functional requirements. Every piece of the input data is not precisely deterministic but may have some fluctuations. However, it does not mean that the deterministic formulation is not adequate. Take, for example, the primary source of uncertainty in many business-related problems—demand. Generally, a stochastic formulation might be overkill in the near-term horizon. Contrarily, a long-term horizon might require consideration of several scenarios for future demand. The rule is not rigid. For example, a famous *newsvendor problem* has a near-term horizon—a day. Still, the demand is uncertain.

In most cases, the first step in the modeling process, even in the case of a stochastic formulation, is a deterministic formulation. The deterministic model is a building block for a more elaborate stochastic formulation. Moreover, solving both deterministic and stochastic problems allow you to evaluate the gains of a more challenging stochastic model.

When discussing stochastic modeling, we consider two methods: stochastic programming (Shapiro et al. 2009) and robust optimization (Ben-Tal and Nemirovski 1999). An example of stochastic optimization for the unit commitment problem is in (Ruiz et al. 2009).

1.11   Avoid Direct Connection of the Optimization Model to a Database

Usually, a database stores all data needed for an optimization model. However, a direct connection of your model to the database can be impractical.

- First, any changes in the database schema affect your code.
- Second, replacing the database with another one ruins all your code.
- Third, if the database goes down, your work will also be on hold.

My solution is intermediate storage. I use CSV (Comma-Separated Values) files for this purpose. It would help to use a query language (such as MySQL for relational databases) to export data from the database as CSV files (Fig. 1.9).

You can quickly analyze CSV files using Microsoft Excel or Google Spreadsheets. You can debug your code autonomously. Moreover, it makes your solution portable: almost any database connects to the CSV files.

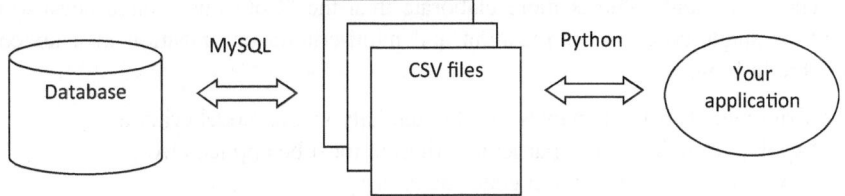

Fig. 1.9 A CSV intermediate storage

Table 1.1 A configuration file example

Constraint	Apply			
InventoryDynamics	1			
QtyProductArrival	1			
QtyProductDeparture	1			
OneArrivalPerCycle	1			
MaxInventoryVolume	1			
MinInventoryVolume	1			
InventoryLowerBound	1			
InventoryUpperBound	1			
ArrivalTimeSpread	1			
Criterion	Apply	Goal	Rank	Tolerance
MinMaxVolume	1	min	1	0.01
MaxMinVolume	1	max	2	0.01
Solver	Apply			
SCIP	1			
Gurobi	0			

1.12   Implement Your Model with Maximum Flexibility

It was beneficial to implement a model not as a rigid set of constraints and objective functions but as manageable entities.

For this purpose, a particular configuration CSV file accompanies every model. In its simple form (see Table 1.1), the configuration file has three tables called "Constraint", "Criterion" and "Solver". For example, please see below a configuration file for a discrete model of the Warehouse Storage Space Problem (Chap. 2).

The "Constraint" table is simple, with two columns:

- "Constraint." It lists the names of all potentially viable model constraints;
- "Apply." It indicates the inclusion (1/0) of every constraint in the model.

The script which reads the configuration file accumulates only included constraints. The result is a set (in Python) of active constraints.

You can code the model's constraints using "switch" or "if-else" statements. For example, see Appendix C.

The "Criterion" table is more elaborate than the "Constraint" table because it controls single-criterion optimization and multi-criteria optimization in a lexicographic fashion.

- "Criterion." It lists the names of all potentially viable model criteria;
- "Apply." It indicates if a particular criterion must be applied (1/0);
- "Goal." It is an optimization goal (max/min);
- "Rank." It is a criterion rank (1, 2, ...);
- "Tolerance." It is an absolute tolerance for every criterion if we impose a bound on it.

The "Solver" table lists available solvers.

- "Solver." It lists all available solvers;
- "Apply." It indicates a particular solver (1/0).

In this case, the optional solvers are SCIP (Optimization Online 2022) and Gurobi (Gurobi Optimizer Reference Manual 2020), and SCIP is picked for the run.

An advantage of the configuration file is no need to change the code if you want to experiment with the model. You adjust the configuration file accordingly. Moreover, even an end-user can play with the model if you create a simple interface with the configuration file or just instruct the user how to manage the CSV file.

1.13   Provide Consistency in the Results

Customers might become frustrated if they run an application today and receive a slightly different solution than the one, they received yesterday. Ensuring a stable and consistent output is paramount in maintaining customer satisfaction. As a practitioner, you should strive for consistency across successive model runs with identical input data.

Discrepancies in outcomes may sometimes derive from the nuances of floating-point arithmetic. For instance, avoid direct numerical comparisons of floating-point numbers like:

$$if(a == b) :$$

because inconsistent and unpredictable results may ensue. Instead, always use a tolerance or threshold:

$$if(abs(a - b)) \leq tolerance :$$

Challenges further arise when your model traverses varying operating systems and platforms. Navigating this intricacy while preserving uniformity in results is a task that underscores the essence of your role.

Chapter 2
Real-World Problems

2.1 Cutting Stock Problem

Though initially stemming from practical requirements, the one-dimensional Cutting Stock Problem (CSP) has evolved into a classical challenge. Nobel Prize laureate L.V. Kantorovich introduced a linear programming formulation for the CSP while developing scheduling solutions for a paper mill in 1939 (Kantorovich 1939). Subsequently, the problem gained prominence across industries like paper, steel, leather, and others, leading to substantial savings in materials through software implementations.

Classical CSP Statement: The task involves a collection of orders comprising finished (small) rolls with specific widths and an unrestricted quantity of stock (large) rolls with a given width. The objective is to determine the minimum number of stock rolls required to be cut to produce the desired quantity of finished rolls.

In the context of paper mill scheduling, CSP is commonly referred to as *trim optimization* (Lavigne 1993). This scenario entails a paper machine continuously producing a paper that winds into wide and large-diameter jumbo rolls. During a single cutting stage, a winding machine (winder) unwinds a jumbo roll, employs knives (slitters) to cut it according to specified customer widths, and then rewinds it into customer-specified diameters through additional cutting.

Technically, the practical CSP exhibits two dimensions: cutting widthwise (across) as per customer-specified widths and cutting lengthwise (along) according to customer-specified diameters. Nevertheless, the ratio between jumbo roll footage and finished roll footage is significant enough that waste incurred from cutting a jumbo roll across (trim loss) outweighs waste from cutting a jumbo roll lengthwise (slab loss). Notably, a specialized winder, known as a bi-winder, can simultaneously manage finished rolls of varying diameters. While bi-winders introduce some

---

The original version of the chapter has been revised: The equations on Pages 72 and 73 have been corrected. A correction to this chapter is available at:
https://doi.org/10.1007/978-3-031-49838-1_3

E. J. Zak, *How to Solve Real-world Optimization Problems*, SpringerBriefs in Operations Research, https://doi.org/10.1007/978-3-031-49838-1_2

consideration of the second dimension, it is of secondary importance. The subsequent discussion centers on a regular winder that consistently produces finished rolls of the same diameter with each cut along, referred to as a "set." A *cutting pattern* combines the widths of finished rolls present in a set.

Practical CSP Statement: Given the width of a jumbo roll and the widths of finished rolls, the objective is to identify operationally viable cutting patterns and their corresponding activities. These activities aim to fulfill customer demands by generating finished rolls and minimizing the number of sets.

Evidently, the problem's practical formulation is more versatile than the classical one. Moreover, two additional applicable criteria emerge. The first criterion involves pattern minimization, significantly complicating the problem in comparison with the classical CSP. The mathematical model to rigorously solve this problem is overcomplicated, leading to the adoption of more straightforward heuristic approaches for approximate solutions.

The second criterion is focused on minimizing the number of slitter moves. A *slitter move* involves horizontally shifting a slitter from one cutting position to another while establishing a new pattern. This criterion significantly expands the scope of the CSP by incorporating factors related to pattern sequencing and the positioning of rolls within each pattern. A more in-depth exploration of the minimization of slitter moves will be presented in subsequent discussions.

In the following sections, I will outline two primary models for addressing the CSP: the bin-packing and pattern-based models.

2.1.1 Classical Bin-Packing Model

Rather than employing a cutting process, the bin-packing model revolves around an item-based packing procedure. The problem can be rephrased as follows: Determine the smallest quantity of identical bins required to pack a given set of items within those bins.

Sets

$I = \{1, 2, \ldots, m\}$ is the set of items.
$K = \{1, 2, \ldots\}$ is the set of identical bins.
Z_+^1 is the set of non-negative integers.

Input Data

w_i is the width of item $i \in I$.
b_i is the total number of items $i \in I$.
w^{bin} is the bin width.

Variables

x_{ik} = number of items $i \in I$ packing in a bin $k \in K$.
y_k = indication of using bin $k \in K$ for packing items.

Model

$$\text{Minimize} \quad \sum_{k \in K} y_k \tag{2.1}$$

$$\text{Subject to} \quad \sum_{i \in I} w_i x_{ik} \leq w^{bin} y_k, \quad k \in K. \tag{2.2}$$

$$\sum_{k \in K} x_{ik} = b_i, \quad i \in I. \tag{2.3}$$

$$x_{ik} \in Z_+^1, \quad i \in I, \quad k \in K. \tag{2.4}$$

$$y_k \in \{0, 1\}, \quad k \in K. \tag{2.5}$$

The criterion (2.1) minimizes the number of used bins.
The constraints (2.2) are the bin width restrictions on the packing items.
The constraints (2.3) force to pack all items.
The bounds (2.4) define the integrality of the number of items.
The bounds (2.5) define binary conditions on the used bins.

Remarks

- Though the set K seems to be infinite, it has an apparent maximum cardinality: $|K| \leq \sum_{i \in I} b_i$. A more accurate upper bound is:

$$|K| \leq \sum_{i \in I} \left\lceil \frac{b_i}{\lfloor w^{bin}/w_i \rfloor} \right\rceil. \tag{2.6}$$

Of course, this cardinality serves as an upper bound of the objective function.
- Sometimes the bin-packing model is formulated so you can see binaries $x_{ik} \in \{0, 1\}$ instead of non-negative integers $x_{ik} \in Z_+^1$. It means that items with identical widths are not grouped, but considered as distinct items. In this case, $b_i = 1, \forall i$.
- To make bin-packing model more suitable for the CSP, we reformulate the equality constraint (2.3) as an inequality:

$$\sum_{k \in K} x_{ik} \geq b_i, \quad i \in I.$$

Let us consider a following example:

Example 2.1 CSP. The input data is in Table 2.1 below.

Table 2.1 Input data

Item	Type	Width, w_i	Number of items, b_i
	Bin	200	–
A	Item	16.5	32
B	Item	17	47
C	Item	18	26
D	Item	23	64
E	Item	25.75	32
F	Item	29	37
G	Item	31.25	22
H	Item	32	15

According to the Formula (2.6) the upper bound is 38 bins, and the optimal solution necessitates the use of 32 bins. Appendix A, Table A.1, provides a detailed breakdown of this solution across all 32 bins. Each bin distinguishes itself from the others based on its assortment of items.

The bin-packing model offers a straightforward representation of the Cutting Stock Problem (CSP). However, it has a notable drawback—the abundance of potential ways to arrange items within bins. This leads to a significant proliferation of distinct cutting patterns. In Table A.1, all 32 bins (sets) have unique item compositions. Paper mill schedulers often prefer a more limited variety of bins (cutting patterns).

Definition The *integrality gap* refers to the absolute difference between the objective function values of the integer (or Mixed Integer Programming—MIP) problem and its Linear Programming (LP) relaxation solution.

In the given example, the objective value of the LP relaxation is 31.6575. Therefore, the integrality gap, which is calculated as $32 - 31.6575 = 0.3425$, is small. Furthermore, as we will observe later, this gap is identical to the pattern-based model.

Theorem 2.1 The integrality gap associated with the bin-packing model is greater than or equal to that associated with the pattern-based model. The proof can be found in Appendix B.

The higher integrality gap complicates the immediate adoption of the bin-packing model for solving CSP. Consequently, the progression leads to a more practical, pattern-based model.

2.1.2   Pattern-Based Model

What is a cutting pattern?

Definition   A *cutting pattern* (or pattern) is a combination of finished items $\mathbf{a} = (a_1, a_2, \ldots, a_m)^T$ such that the total width cannot exceed the stock item width w^{stock}, i.e.

$$\sum_{i \in I} w_i a_i \leq w^{stock}.$$

Definition   *Pattern width* is the total width occupied by finished items in the pattern:

$$\sum_{i \in I} w_i a_i.$$

The theoretical upper bound of the pattern set J cardinality is:

$$|J| \leq \frac{1}{m!} \left(\left\lfloor w^{stock} / \min_i \{w_i\} \right\rfloor + 1 \right)^m \qquad (2.7)$$

Let us consider the following example:

Example 2.2   CSP. The data set is defined in Table 2.2.

For the above dataset, the upper bound calculated by the Formula (2.7) is 36, while the number of cutting patterns is 22. The list of all cutting patterns is in Table 2.3:

Of course, Example 2.2 is a toy example. In more practical examples, the number of patterns might be huge. Let us consider Example 2.3 that essentially is the same as for bin-packing data set in Table 2.1.

Example 2.3   CSP. The data set defined in Table 2.4.

In this example, the upper bound of the total number of patterns calculated by the Formula (2.7) is 20,231 vs. the actual number of 14,898. I will show a solution to the corresponding CSP later.

Table 2.2 Input data

#	Roll type	Width
	Stock	10
1	Finished	2
2	Finished	3
3	Finished	4

Table 2.3 Cutting patterns

#	Pattern	Pattern width
1	[0 0 1]	4
2	[0 0 2]	8
3	[0 1 0]	3
4	[0 1 1]	7
5	[0 2 0]	6
6	[0 2 1]	10
7	[0 3 0]	9
8	[1 0 0]	2
9	[1 0 1]	6
10	[1 0 2]	10
11	[1 1 0]	5
12	[1 1 1]	9
13	[1 2 0]	8
14	[2 0 0]	4
15	[2 0 1]	8
16	[2 1 0]	7
17	[2 2 0]	10
18	[3 0 0]	6
19	[3 0 1]	10
20	[3 1 0]	9
21	[4 0 0]	8
22	[5 0 0]	10

Table 2.4 Input data

Item	Roll type	Width, w_i	Ordered amount, b_i
	Stock	200	–
A	Finished	16.5	32
B	Finished	17	47
C	Finished	18	26
D	Finished	23	64
E	Finished	25.75	32
F	Finished	29	37
G	Finished	31.25	22
H	Finished	32	15

Number of Sets Minimization

Let us assume that we generate all feasible patterns in the preprocessing step. We come up with the following pattern-based model $CPB(A, \mathbf{b})$:

$$Minimize \quad \mathbf{1}^T \mathbf{x}$$

$$\textit{Subject to}\quad A\mathbf{x} \geq \mathbf{b}\quad CPB(A, \mathbf{b})$$

$$\mathbf{x} \in Z_+^n$$

Here, we use a matrix notation:

$\mathbf{1} = (1, 1, \ldots, 1)^T$ is a vector of n dimension.

A is a matrix of $m \times n$ where every column $\mathbf{a}_j = \left(a_{1j}, a_{2j}, \ldots, a_{mj}\right)^T$, $j \in J = \{1, 2, \ldots, n\}$ is a pattern.

$\mathbf{b} = (b_1, b_2, \ldots, b_m)^T$ is a vector of order amounts.

$\mathbf{x} = (x_1, x_2, \ldots, x_n)^T$ is a vector of variables which defines pattern activities.

Z_+^n is a set of n-dimensional nonnegative integers.

The objective function minimizes the total pattern activities, equivalent to the number of sets (or used stock items if they have the same diameter as finished rolls).

Theorem 2.2 If the matrix in the model $CPB(A, \mathbf{b})$ encompasses all feasible patterns, the bin-packing and pattern-based models constitute alternative formulations of the same problem.

This equivalence signifies that both formulations yield identical objective function values. You can find the supporting proof in Appendix B.

Addressing the pattern-based model presents various methods. Given the potentially vast number of feasible patterns leading to numerous integer variables, the direct solution of this Integer Programming (IP) model can be challenging. Practitioners often employ the following approach: they tackle an LP relaxation of the model $CPB(A, \mathbf{b})$ and employ successive rounding. While achieving an optimal solution is not assured, this strategy performs well in most practical instances.

An inherent concern arises when the number of potential patterns becomes exceedingly large, and the LP relaxation struggles to manage this matrix configuration. There are two intelligent strategies to mitigate this challenge:

Offline Column Generation: A limited set of patterns can be intelligently selected to form a matrix A a priori. This approach, known as offline pattern generation, can curtail the complexity.

Dynamic Column Generation: An ingenious dynamic column generation technique, as proposed by P.C. Gilmore and R.E. Gomory (1961, 1963), offers a powerful alternative. This technique is one of the most captivating algorithms in Operations Research and holds substantial practical value.

Let us start with the offline column generation. It is crucial to consider a customer's multiple functional requirements for the patterns. Below is a compilation of frequently encountered functional prerequisites:

Limitation on finished item count: This constraint stems from the finite number of knives (slitters) available on the winder machine for cutting jumbo rolls. Consequently, the number of finished items in a pattern is restricted.

Controlled variety of roll widths: Patterns are bound by limitations regarding the variety of finished roll widths. This restriction arises from downstream operations such as loading.

Restriction on "big" roll count: To account for weight limitations on the back stand, patterns are constrained in terms of the number of "big" rolls they can include.

Avoidance of close sizes: Patterns are designed to prevent the inclusion of rolls with closely similar sizes. This consideration facilitates visual identification by the crew, eliminating the need for constant measurements.

While incorporating such requirements through simple filters during pattern generation offers a relatively straightforward approach, it is essential to acknowledge that this might impact the performance of pattern generation. Balancing these practical needs with computational efficiency is a vital consideration. Notably, these functional constraints are pivotal in simplifying the problem by significantly reducing the pool of feasible patterns. For instance, constraining patterns in Example 2.3 to a maximum of 8 finished rolls results in a reduction from 14,898 to 10,233 feasible patterns.

However, the need for more sophisticated routines becomes evident. For instance, certain requirements necessitate a more intricate implementation of a knapsack algorithm. This development falls under the purview of responsibilities for an Operations Research practitioner.

Now, we will consider the online (dynamic) column generation.

The following is an LP relaxation of the pattern-based model and its corresponding dual counterpart.

Primal	Dual
Minimize $\mathbf{1}^T \mathbf{x}$	Maximize $\mathbf{b}^T \mathbf{u}$
Subject to $A\mathbf{x} \geq \mathbf{b}$	Subject to $A^T \mathbf{u} \leq \mathbf{1}$
$\mathbf{x} \geq 0$	$\mathbf{u} \geq 0$

Here, $\mathbf{u} = (u_1, u_2, \ldots, u_m)^T$ is a vector of dual variables. We can treat every dual variable u_i as a price of roll $i \in I$ of width w_i and length 1 relative to the cost of the stock roll of width w^{stock} and length 1. The dual model maximizes the total revenue subject to the cost of every pattern less or equal 1.

The strong duality tells us that in the optimal solution:

$$\mathbf{1}^T \mathbf{x}^{opt} = \mathbf{b}^T \mathbf{u}^{opt}.$$

Every constraint in the dual problem has a form:

$$\mathbf{a}^T \mathbf{u} \leq 1,$$

where vector $\mathbf{a} = (a_1, a_2, \ldots, a_m)^T$ is a pattern.

Given the vector of dual variables $\mathbf{u}$, which comes from solving the primal problem, we devise a knapsack problem where a pattern $\mathbf{a}$ is a variable.

$$\textit{Maximize} \quad \mathbf{u}^T\mathbf{a}$$

$$\textit{Subject to} \quad \mathbf{w}^T\mathbf{a} \leq w^{stock} \quad KS\left(\mathbf{u}, \mathbf{w}, w^{stock}\right)$$

$$\mathbf{a} \in Z_+^m$$

Solving this classical knapsack $KS(\mathbf{u}, \mathbf{w}, w^{stock})$, we will get an optimal value of the objective function $\mathbf{u}^T\mathbf{a}$. If this value $\mathbf{u}^T\mathbf{a} \leq 1$, then we found a solution for the LP relaxation of the pattern-based model. The primal and the dual problem are in the proper relationship.

If this value $\mathbf{u}^T\mathbf{a} > 1$, then the pattern $\mathbf{a}$ is missed in the matrix A. We must add this pattern to the matrix A and solve the primal problem again.

If you can implement a *revised simplex*, there is no need to solve the primal problem from scratch, and you do not need to use a third-party solver. In this case, you can overload the column selection, so along with testing the non-basic columns for entering the basis, the knapsack problem $KS(\mathbf{u}, \mathbf{w}, w^{stock})$ solution also becomes a potential candidate for entering the basis.

On the other hand, should you choose to depend on a third-party solver, your programming task becomes more straightforward. The accompanying flowchart outlines the comprehensive procedure (Fig. 2.1).

The Gilmore-Gomory dynamic columns generation finds an optimal solution of the LP relaxation of CSP.

Example 2.3 (cont.) Solving an LP relaxation of this example gives the optimal value of the functional equal to 31.6575. The table below (Table 2.5) presents an LP solution. As we can see, this solution is a perfect match for the ordered amount and has no trim loss.

What is interesting is that the LP objective value matches the lower bound of the optimal number of sets calculated as:

$$LowerBound = \mathbf{w}^T\mathbf{b}/w^{stock}. \tag{2.8}$$

It means the LP solution has no trim loss since all patterns in the solution have the maximum possible width equal to w^{stock}. However, the sets are not integers. So, an additional step is needed to find an integer solution.□

An integer programming problem is said to have the *integer round-up property* if its optimal value does not exceed the least integer greater than or equal to its linear LP relaxation value, i.e.

$$optimalValue \leq \lceil LPoptimalValue \rceil.$$

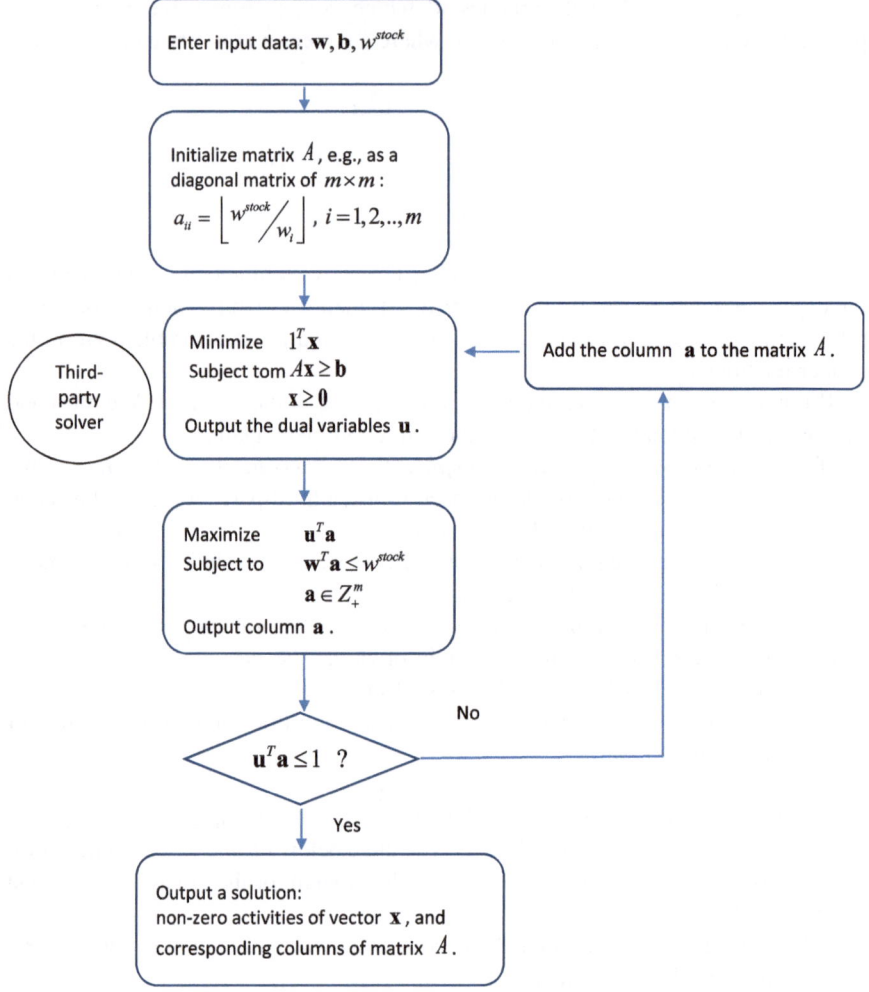

Fig. 2.1 A simplified Gilmore-Gomory dynamic columns generation

O. Marcotte (1985) proved that certain classes of cutting stock problems have the integer round-up property. Most of the practical problem falls into the valid round-up property class. There is another way to formulate this property.

For most practical CSPs, the integrality gap is less than 1.

$$| optimalValue - LPoptimalValue | \leq 1.$$

There are few known instances of CSP with an integrality gap of 1 or more. There are no known instances with an integrality gap more or equal 2. Open conjecture: An integrality gap for any CSP is less than 2 (Scheithauer and Terno 1995; Zak 2006). In special instances of CSPs, the conjecture was proved to be true (Martinovich 2022).

Table 2.5 LP solution of Example 2.3

Item	Sets	Patterns								Sch	Ord	O/U
		1	2	3	4	5	6	7	8			
		8.266	8.894	2.26	6.998	0.894	0.967	1.062	2.311	31.6575		
	Widths									Sch	Ord	O/U
A	16.5	0	0	6	2	0	0	2	1	32	32	0
B	17	1	4	1	0	1	0	0	0	47	47	0
C	18	2	0	0	0	0	1	8	0	26	26	0
D	23	0	2	1	6	1	0	1	0	64	64	0
E	25.75	1	1	0	0	0	1	0	6	32	32	0
F	29	2	1	1	1	0	0	0	1	37	37	0
G	31.25	1	1	0	0	0	5	0	0	22	22	0
H	32	1	0	1	0	5	0	0	0	15	15	0
	Pattern width	200	200	200	200	200	200	200	200			

We will demonstrate later that the optimal value of the objective function in Example 2.3 is 32. It means that the rounding-up property holds for this instance: $\lceil 31.6575 \rceil = 32$.

We can use this round-up property during rounding of a non-integer solution.

$$\text{Minimize} \quad \mathbf{w}^T \left\| A\mathbf{x} - \mathbf{b} \right\|$$

$$\text{Subject to} \quad \mathbf{1}^T\mathbf{x} = s$$

$$\mathbf{x} \in Z_+^m$$

where

A is a solution matrix generated by the dynamic column generation.

s is the rounded up the optimal number of sets found by the column generation. Here, we allow a deviation from the ordered amount $\mathbf{b}$.

If we use L^1 norm of vector $A\mathbf{x} - \mathbf{b}$, then we can linearize the model:

$$\text{Minimize} \quad \mathbf{w}^T(\mathbf{u} + \mathbf{v})$$

$$\text{Subject to} \quad A\mathbf{x} - \mathbf{u} + \mathbf{v} = \mathbf{b}$$

$$\mathbf{1}^T\mathbf{x} = s$$

$$\mathbf{x} \in Z_+^m$$

$$\mathbf{u}, \mathbf{v} \in R_+^m$$

The latest is an MIP problem solved by a generic optimization solver.

Example 2.3 (cont.). Solving the rounding problem gives an integer solution (Table 2.6) with the objective function value of 137.5.

Table 2.6 LP-based solution of Example 2.3

Item		Patterns										
		1	2	3	4	5	6	7	8			
Item	Sets	8	9	2	7	1	1	1	3	32		
	Widths									Sch	Ord	O/U
A	16.5	0	0	6	2	0	0	2	1	31	32	−1
B	17	1	4	1	0	1	0	0	0	47	47	0
C	18	2	0	0	0	0	1	8	0	25	26	−1
D	23	0	2	1	6	1	0	1	0	64	64	0
E	25.75	1	1	0	0	0	1	0	6	36	32	4
F	29	2	1	1	1	0	0	0	1	37	37	0
G	31.25	1	1	0	0	0	5	0	0	22	22	0
H	32	1	0	1	0	5	0	0	0	15	15	0
	Pattern width	200	200	200	200	200	200	200	200			

The solution is perfect regarding trim loss and the number of sets (round-up of the LP objective function value) but does not match precisely the ordered amount. Strongly speaking, this solution is not a solution of the original problem since it violates the ordered amount for items A and C. However, it is not the main reason why this solution is not practical. A notable drawback of this solution is its high count of cutting patterns. Developing fewer patterns proves challenging when utilizing an LP approach, where the number of columns in a non-degenerate scenario equals the number of rows. Subsequently, I will showcase a heuristic solution to the same problem that boasts half the number of patterns compared to the LP solution.□

Remark The LP value surpasses the lower bound value in numerous CSPs. The ensuing simple example serves to illustrate this phenomenon.

Example 2.4 CSP. This example includes a single finished roll (Table 2.7).

The lower bound is $35 \times 99/200 = 17.325$. A feasible pattern is $\mathbf{a} = (5)$. The current pattern exhibits poor performance due to a significant trim loss. Unfortunately, given the nature of the problem, there is no feasible alternative that yields better results. Upon using an LP solver, the optimal activity level for the pattern equals $99/5 = 19.8$ sets.

A notable disparity is evident between the lower bound (17.325) calculated by Formula (2.8) and the optimal value determined by the LP solver (19.8). The rounding-up characteristic is also applicable in this scenario, leading to an optimal value of 20 within the integer program: $\lceil 19.8 \rceil = 20$.

However, the solution (Table 2.8) still leaves us with an excess of one roll of 35 inches.

By the way, this instance confirms our earlier statement (Theorem 2.1) that the integrality gap associated with the bin-packing model is greater than or equal to that associated with the pattern-based model. The integrality gap for the bin-packing problem is $20 - 17.325 = 2.675$, whereas for the pattern-based model, it is only $20 - 19.8 = 0.2$.□

Table 2.7 Input data

#	Roll type	Width	Ordered amount
	Stock	200	–
1	Finished	35	99

Table 2.8 Solution of Example 2.4

		Pattern			
		1			
Item	Sets	20	20		
	Widths		Sch	Ord	O/U
A	35	5	100	99	1
	Pattern width	175			

Trim Loss Minimization Model

In contrast to the previously discussed approach of minimizing the total number of sets, the trim loss minimization model within the CSP focuses directly on reducing trim losses. This approach becomes particularly relevant when we define a specified range for ordered quantities or when an application employs what is known as "help rolls".

Definition A *help roll* is an auxiliary finished roll not ordered by any customer but proceeds directly to storage for future use. Incorporating help rolls into the process can enhance operational efficiency and decrease trim losses.

Now, let us proceed by formally defining both a "pattern trim loss" and a "total trim loss":

Definition A *pattern trim loss* refers to the portion of unused width in a stock item:

$$w^{stock} - \mathbf{w}^T \mathbf{a}.$$

The *total trim loss* is the summation of trim loss across all patterns:

$$\left(\mathbf{1}^T w^{stock} - \mathbf{w}^T A\right)\mathbf{x}.$$

The table below shows a couple of primal and dual problems in the case of the LP relaxation of the total trim loss minimization.

Primal	Dual
Minimize $(\mathbf{1}^T w^{stock} - \mathbf{w}^T A)\mathbf{x}$	Maximize $(\mathbf{b}^{min})^T \mathbf{u} - (\mathbf{b}^{max})^T \mathbf{v}$
Subject to $\mathbf{b}^{max} \geq A\mathbf{x} \geq \mathbf{b}^{min}$	Subject to $A^T(\mathbf{u} - \mathbf{v}) \leq \mathbf{1} w^{stock} - A^T \mathbf{w}$
$\mathbf{x} \geq 0$	$\mathbf{u}, \mathbf{v} \geq 0$

Remark If an application uses help rolls, the corresponding constraints look like:

$$b_i^{max} \geq \sum_j a_{ij}x_j, \quad i \in I^{help},$$

where
I^{help} = subset of help rolls, $I^{help} \subset I$.
b_i^{max} = the maximum allowable number of help rolls $i \in I^{help}$.
A question may arise—why ordered quantities must be in a range here instead of a point as in the earlier patterns-based model? The answer is to prevent a big overrun. The following example illustrates this answer.

Example 2.5 CSP. Let us consider the data set in Table 2.4 a minor modification: the stock roll size instead of 200 has a width of 192.5. The input data is in Table 2.9.

If you formulate the problem with one-sided constraints, namely:

Table 2.9 Input data

i	Roll type	Width, w_i	Ordered amount, b_i
	Stock	192.5	–
A	Finished	16.5	32
B	Finished	17	47
C	Finished	18	26
D	Finished	23	64
E	Finished	25.75	32
F	Finished	29	37
G	Finished	31.25	22
H	Finished	32	15

Table 2.10 Solution of Example 2.5 with one-sided constraints

		Patterns			
		1			
Item	Sets	64	**64**		
	Widths		Sch	Ord	O/U
1	16.5	1	64	32	32
2	17	1	64	47	17
3	18	1	64	26	38
4	23	1	64	64	0
5	25.75	1	64	32	32
6	29	1	64	37	27
7	31.25	1	64	22	42
8	32	1	64	15	49
Pattern width		192.5			

$$\text{Minimize} \quad \left(\mathbf{1}^T w^{stock} - \mathbf{w}^T A\right)\mathbf{x}$$

$$\text{Subject to} \quad A\mathbf{x} \geq \mathbf{b}$$

$$\mathbf{x} \geq 0$$

then we devise an obvious solution even manually—Table 2.10:

The solution is a single pattern $\mathbf{a} = (1, 1, \ldots, 1)^T$ of 64 sets and no trim loss. The drawback is a massive overrun of 237 rolls. This example confirms the need of the two-sided formulation.

$\square$

Every constraint in the dual problem has a form:

$$\mathbf{a}^T (\mathbf{u} - \mathbf{v}) \leq w^{stock} - \mathbf{a}^T \mathbf{w}$$

A corresponding knapsack problem looks like:

$$Maximize \quad (\mathbf{u} - \mathbf{v} + \mathbf{w})^T \mathbf{a}$$

$$Subject\ to \quad \mathbf{w}^T \mathbf{a} \leq w^{stock} \quad KS(\mathbf{u} - \mathbf{v} + \mathbf{w}, \mathbf{w}, w^{stock})$$

$$\mathbf{a} \in Z_+^m$$

In the revised simplex implementation, along with testing the non-basic columns, you should solve a knapsack problem $KS(\mathbf{u} - \mathbf{v} + \mathbf{w}, \mathbf{w}, w^{stock})$. Its solution $\mathbf{a}$ becomes a potential candidate to enter the basis.

If you use a third-party solver, the overall scheme is the same as in the classical dynamic column generation except the checking block (see Fig. 2.2).

Both trim loss minimization and number of sets minimization models share a common limitation. Due to the matrix's square nature, the count of patterns within non-degenerate solutions corresponds to the count of items. For instance, if there are 20 items, the resulting solution would naturally consist of 20 patterns. However, some gurus in the paper mill industry can devise solutions involving just eight patterns—a seemingly implausible scenario. This apparent contradiction arises from degeneracy within Linear Programming (LP).

While there are specialized models designed for minimizing the count of patterns, their scalability poses a concern. A more effective approach emerges in pursuing a practical and scalable alternative, which we will explore in the upcoming section.

Sequential Heuristic Procedure

In addition to the various compelling reasons for developing a heuristic algorithm mentioned in Chap. 1, it is crucial to emphasize another one specific to CSP: solution quality. We aim for procedures that produce a significantly smaller number of patterns compared to the LP approach.

In this context, I present a variant of the heuristic originally devised by R.W. Haessler (1968) for the CSP. I intend to establish certain generalizations tailored to practical formulations of the CSP.

Initially, I will lay out a principle of optimality relevant to the CSP.

Principle of optimality

If $\mathbf{x} = (x_1, x_2, x_3, \ldots, x_n)^T$ is an optimal solution of the pattern-based model $CPB(A, \mathbf{b})$, then the vector $(x_1, \ldots, x_{j-1}, x_{j+1} \ldots, x_n)^T$ is an optimal solution of the reduced model $CPB(A \backslash \mathbf{a}_j, \mathbf{b} - \mathbf{a}_j x_j)$, where $\mathbf{a}_j$ is jth column of the matrix A; $A \backslash \mathbf{a}_j$ is the matrix A without column $\mathbf{a}_j$.

Original Haessler's heuristic does not implement Bellman's dynamic programming algorithm (Bellman 1957). However, it explores the concept of generating sequential patterns. I blend the concept of sequential pattern generation and dynamic programming algorithm into the following recursive procedure called the Sequential Heuristic Procedure (SHP), presented in Fig. 2.3.

Now the heuristic run looks like in Fig. 2.4.

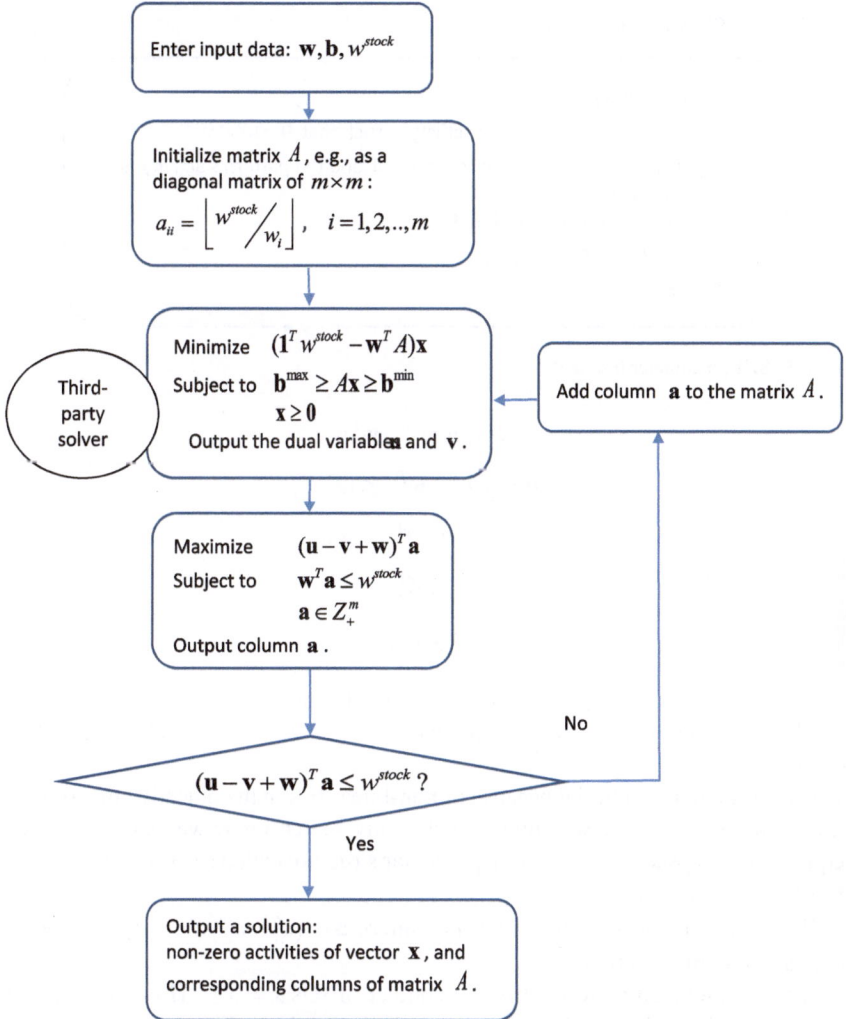

Fig. 2.2 A simplified dynamic column generation for trim loss minimization

The most difficult part of $SHP(\mathbf{w}, \mathbf{d}, w^{stock}, \alpha)$ is "Step 2". Two typically conflicting tendencies simultaneously contribute to the same objective function:

- Select a pattern with the widest possible width.
- Choose the number of corresponding sets to be as large as possible.

We can formalize "Step 2" of $SHP(\mathbf{w}, \mathbf{d}, r, \alpha)$ as follows:

$SHP(\mathbf{w}, \mathbf{d}, w^{stock}, \alpha)$

1. If $\mathbf{d} = \mathbf{0}$ go to End.
2. Find a pattern $\mathbf{a}$ and its activity x such that $\mathbf{d} - \mathbf{a}x \geq 0$.
3. Include the pattern $\mathbf{a}$ into matrix A and the pattern's activity x into vector $\mathbf{x}$.
4. Update vector $\mathbf{d}$ as $\mathbf{d} := \mathbf{d} - \mathbf{a}x$.
5. $SHP(\mathbf{w}, \mathbf{d}, w^{stock}, \alpha)$
6. End.

Fig. 2.3 SHP; parameter $0 < \alpha \leq 1$

$$Maximize \quad x\mathbf{w}^T\mathbf{a}.$$

$$Subject\ to \quad \mathbf{w}^T\mathbf{a} \leq w^{stock},$$

$$\mathbf{a}x \leq \mathbf{d},$$

$$x \in Z_+^1,$$

$$\mathbf{a} \in Z_+^m.$$

The concept behind this model follows a greedy approach: each generated pattern and the associated number of sets attempt to maximize the removal of the ordered amount.

This problem is not trivial because of non-linearities in the objective function and constraints. However, if we apply a dichotomy search on x, we can solve it as a sequence of knapsack problems of a particular structure called a subset-sum problem (Kellerer et al. 2004).

The dichotomy search ends with the pattern $\mathbf{a} = \left(a_1^{opt}, a_2^{opt}, \ldots, a_m^{opt}\right)^T$, and the optimal value of its activity x^{opt}.

There is a temptation to finalize the number of sets $x = x^{opt}$. However, ignoring downstream difficulties makes this approach shortsighted. That is why $SHP(\mathbf{w}, \mathbf{d}, w^{stock}, \alpha)$ has a parameter α, $0 < \alpha \leq 1$, for picking a pattern activity between 1 and $\lfloor \alpha x^{opt} \rfloor$.

$$x = \max\{1, \lfloor \alpha x^{opt} \rfloor\}$$

This approach makes the heuristic less "greedy" and more efficient.

Example 2.3 (cont.)
A heuristic solution to this problem is in Table 2.11:

This represents an optimal solution to the original problem—Example 2.3, as the objective function value (32) is the rounded-up result of the LP relaxation (31.6575).

In this context, a pattern index also serves as a sequential number for pattern generation. One notable advantage of this solution is its minimal number of patterns,

Fig. 2.4 SHP run

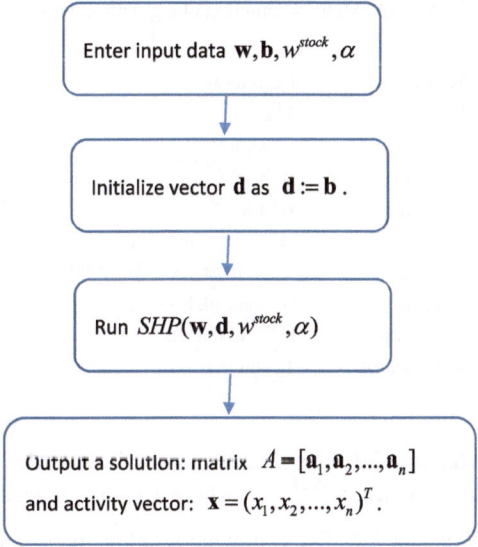

Enter input data $\mathbf{w}, \mathbf{b}, w^{stock}, \alpha$

Initialize vector $\mathbf{d}$ as $\mathbf{d} := \mathbf{b}$.

Run $SHP(\mathbf{w}, \mathbf{d}, w^{stock}, \alpha)$

Output a solution: matrix $A = [\mathbf{a}_1, \mathbf{a}_2, ..., \mathbf{a}_n]$ and activity vector: $\mathbf{x} = (x_1, x_2, ..., x_n)^T$.

Table 2.11 Heuristic solution of Example 2.3

		Patterns							
		1	2	3	4				
Item	Sets	16	11	3	2	**32**			
	Widths					Sch	Ord	O/U	
A	16.5	0	0	6	7	32	32	0	
B	17	0	4	1	0	47	47	0	
C	18	1	1	0	0	27	26	1	
D	23	4	0	0	0	64	64	0	
E	25.75	0	2	2	2	32	32	0	
F	29	2	0	1	1	37	37	0	
G	31.25	0	2	0	0	22	22	0	
H	32	1	0	0	0	16	15	1	
Pattern width		200	200	196.5	196				

totaling only four. This is half the number generated by LP-based solutions (Table 2.6). However, a drawback is the utilization of patterns that incur some trim loss, specifically Patterns #3 and #4.

Metrics

The table below (Table 2.12) provides a comparison of all solutions for the same trim problem presented in Example 2.3, using a set of widely recognized metrics.

Table 2.12 Metrics definition along with specific values for Example 2.3

Metric	Definition	Example 2.3		
		LP solution	IP solution	Heuristic
Input	$\mathbf{1}^T\mathbf{x}$	31.6575	32	32
Output	$\mathbf{w}^T A\mathbf{x}/w^{stock}$	31.6575	32	31.9075
OrdQty	$\mathbf{w}^T\mathbf{b}/w^{stock}$	31.6575	31.6575	31.6575
LowerBound	$\mathbf{w}^T\mathbf{b}/w^{stock}$	31.6575	31.6575	31.6575
TrimLoss	$(\mathbf{1}^T - \mathbf{w}^T A/w^{stock})\mathbf{x}$	0.0	0.0	0.0925
OverRun	$\mathbf{w}^T \max{(\mathbf{0}, A\mathbf{x} - \mathbf{b})}/w^{stock}$	0.0	0.515	0.25
UnderRun	$\mathbf{w}^T \max{(\mathbf{0}, \mathbf{b} - A\mathbf{x})}/w^{stock}$	0.0	0.1725	0.0
Efficiency	$w^{stock}\mathbf{1}^T\mathbf{x}/\mathbf{w}^T A\mathbf{x}$	100.00%	100%	99.71%
NumPatterns	$\mathbf{1}^T \operatorname{sgn}{(\mathbf{x})}$	8	8	4

Remark 2.1 The measure units of all metrics except Efficiency and NumPatterns is the set size: $w^{stock} \times 1$, where 1 is the reduced roll length.

Remark 2.2 The following relations between metrics are valid:

$$Input = Output + TrimLoss$$

$$Output = OrdQty + OverRun - UnderRun$$

$$LowerBound = OrdQty$$

$$Efficiency = \frac{Output}{Input} 100\%$$

And a composite equation:

$$Input = OrdQty + OverRun - UnderRun + TrimLoss$$

Differences Between the Classical and Real-World CSP

The real-world CSP differs from the classical one by numerous additional functional requirements. Table 2.13 illustrates the differences.

2.1.3   Conclusions

The cutting stock problem involves efficiently converting jumbo rolls into customer-specified finished rolls. The CSP exhibits various interpretations depending on the industry context.

Addressing customer-specific functional requirements is essential to tackle this practical issue. These functional requirements play a dual role. Certain requirements

Table 2.13 Functional requirements for the classical and a real-world CSP

Classical CSP	Real world CSP
Ordered amount must be satisfied.	Ordered amount must be approximately satisfied (MAJIQTRIM: Trim optimization software 1995).
The stock item width is given.	The stock roll width can slightly vary depending on the winder speed (MAJIQTRIM: Trim optimization software 1995).
	Multiple stock rolls with different widths (MAJIQTRIM: Trim optimization software 1995).
	The number of finished rolls in a pattern must be restricted (MAJIQTRIM: Trim optimization software 1995).
	The number of small finished rolls in a pattern must be restricted (MAJIQTRIM: Trim optimization software 1995).
	The number of large finished rolls in a pattern must be restricted (MAJIQTRIM: Trim optimization software 1995).
	Number of open lots must be restricted (MAJIQTRIM: Trim optimization software 1995).
	Including loss along (slab loss) with loss across (trim loss) (MAJIQTRIM: Trim optimization software 1995); trimming with a bi-winder.
	Trimming with "help" rolls (MAJIQTRIM: Trim optimization software 1995).
	Parallel running machines—winders (MAJIQTRIM: Trim optimization software 1995).
	Trimming around "bad" spots (MAJIQTRIM: Trim optimization software 1995).
	Elimination of small set patterns (MAJIQTRIM: Trim optimization software 1995).
	Trimming around "dead" zone (patterns in the certain width range are forbidden) (MAJIQTRIM: Trim optimization software 1995).
	Two-stage cutting (see below).
	Trimming in the just-in-time environment (Jonson et al. 1999; MAJIQTRIM: Trim optimization software 1995).

lead to the development of more intricate models, while others narrow down the solution space, enhancing the overall scalability of the formulation.

Diverse models add depth to your modeling toolkit and empower you to examine the solution space from various perspectives. Pattern-based models, LP-based approaches, and heuristics should all find a place in a practitioner's arsenal for solving the CSP. While LP-based models are particularly adept at minimizing trim loss, heuristics prove invaluable for minimizing the number of required patterns.

The CSP inherently involves multiple criteria, as illustrated in Table 2.12. As highlighted in Chap. 1, lexicographic optimization is effective due to the ranking of criteria that stems from the problem's inherent nature.

2.2   Slitter Moves Minimization Problem

The Slitter Moves Minimization Problem (SMMP) frequently constitutes the second phase of many real-world CSP scenarios. It formulates: Given cutting patterns, determine an optimal pattern sequence and roll positioning within each pattern to minimize the number of required slitter moves.

One might even conceptualize the SMMP as a challenge separate from the CSP, encompassing all tasks such as pattern generation, pattern sequencing, and roll positioning. However, achieving this level of complexity is exceptionally ambitious, even within the realm of academic research.

Definition   A *slitter move* is moving a slitter (a cutting knife) to a new position to cut the subsequent pattern. Lifting a slitter to rest is not counted as a slitter move.

The count of necessary slitters to cut a pattern comprising n rolls stand at $n + 1$. Nevertheless, by establishing the position of the left-most slitter as a fixed point, we can simplify matters and assume that the count of potentially "mobile" slitters required to cut n rolls align with the actual number of rolls, thus n.

Here is an example of counting slitter moves.

Example 2.6   SMMP. In this example: $w_1 + w_2 = w_3$ (Fig. 2.5)

There are five slitters and four rolls in two successive patterns. Suppose the upper pattern is the first; the number of slitter moves to cut rolls according to that pattern is four (assuming the left-most slitter #1 does not move). The number of slitters moves to cut the second pattern is also four.

In a similar situation but with more available slitters, the number of slitter moves is one move less (Fig. 2.6).

The total number of available slitters is 6. The positions of slitters allow to lift slitter #2 (which does not account for a move), and slitter #6 goes down from a reserved position. Since slitter #3 keeps its position, switching from Pattern 1 to Pattern 2 requires three slitter moves only.

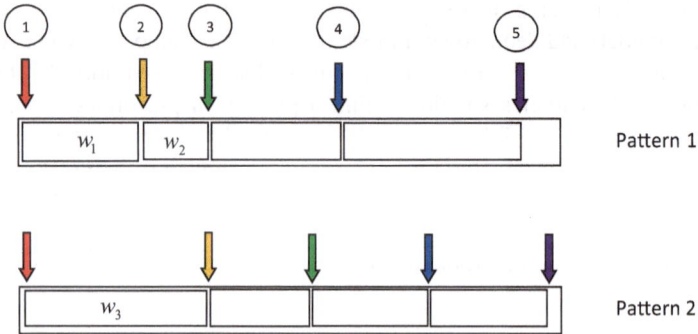

Fig. 2.5 Slitters "starvation", $w_1 + w_2 = w_3$; the number of slitter moves = 4

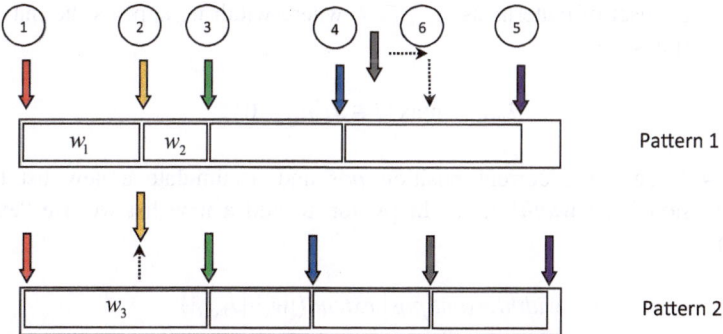

Fig. 2.6 Unrestricted number of slitters; $w_1 + w_2 = w_3$; the number of slitter moves $= 3$

To simplify a further consideration, let us assume we have the second situation with enough slitters.

2.2.1 Heuristic Algorithm

Step 0. Specify sets:

$I = \{1, 2, \ldots, m\}$ is a set of non-identical widths, $w_i \neq w_k$, $i \neq k$.
$J_0 = \{1, 2, \ldots, n\}$ is a set of solution-patterns.

Designate integers:

a_{ij} is the number occurrences of width w_i in the pattern $j \in J_0$.

Step 1. Initialize a two-dimensional list *widthLayout* and a one-dimensional list *patterns*.
Define *npos* as the maximal number of rolls across all patterns. In other words, it is the maximum number of positions.

$$npos = \max_{j \in J_0} \sum_{i \in I} a_{ij}$$

Step 2. Define *tasks* as a stack of all tuples, where every tuple is a starting position *start* of the rolls and a subset of patterns $J \subset J_0$: $(start, J)$.
The starting position *start* is in the range: $1 \leq start \leq npos$.
Initialize *tasks* with a tuple $(1, J_0)$. It means all rolls positions start with position 1 and all patterns are available.
Step 3. Pop (take out) the last tuple $(start, J)$ from the *tasks*.
Step 4. Loop on positions: from $pos = start$ to $pos = npos$ with $step = 1$.
Step 4.1. Pick the width w_k that appears in the maximum number of patterns J.

Define a subset of patterns as $J_{next} \subset J$, where width w_k appears the maximum number of times.

$$J_{next} = \max\{j \in J_k | a_{kj} > 0\}.$$

Step 4.2. Take the current position *pos* and accumulate a new list to the two-dimensional list *widthLayout*. In python to add a new list we use "extend" function:

$$widthLayout[pos].extend([w_k] \cdot |J_{next}|)$$

It means we have added a new list comprised by $|J_{next}|$ identical rolls of width of w_k.

Step 4.3. Conditionally add a new element to the list *patterns*. The condition is the following:

$$\text{If } |J_{next}| = 1 \text{ and } J_{next} = \{j_{next}\}, \text{ then } patterns.append(j_{next}).$$

Step 4.4. Remove a single occurrence of width w_k from all patterns where this width w_k appears:

$$a_{kj} := a_{kj} - 1, \quad j \in J_{next}.$$

Step 4.5. Define the remaining subset of patterns H where width w_k does not appear: $H = J \setminus J_{next}$; redefine $J = J_{next}$. If $H \neq \varnothing$, then push a new tuple (pos, H) to the *tasks* stack.

Step 5. If the *tasks* stack is non-empty, then go to Step 3.

End. Output *widthLayout, patterns*.

Let us illustrate the algorithm on our ongoing Example 2.3.

Example 2.3 (cont.)

We take patterns from the heuristic solution (Table 2.11). Below we designate patterns Pat_1, Pat_2, Pat_3, and Pat_4 as 1, 2, 3, and 4, respectively.

The set of widths $I = \{A, B, C, D, E, F, G, H\}$.

The set of patterns $J_0 = \{1, 2, 3, 4\}$.

Matrix $A = \|a_{ij}\|$, based on patterns $1 - 4$, is in Table 2.14.

The heuristic creates the following solution (Fig. 2.7)

The total number of slitter moves is calculated as $10 + 1 + 7 + 7$, which equals 25. The maximum number of slitter moves is determined, assuming every pattern requires all slitters to move. This calculation results in $10 + 10 + 9 + 8$, totaling 37 moves. Therefore, we can observe a substantial reduction in the number of slitter moves by applying the heuristic mentioned above, going from 37 moves down to 25 moves.

Table 2.14 Matrix $A = \|a_{ij}\|$ is in the highlighted rectangle (see Table 2.11)

		Patterns			
		1	2	3	4
Item	Widths				
A	16.5	0	0	6	7
B	17	0	4	1	0
C	18	1	1	0	0
D	23	4	0	0	0
E	25.75	0	2	2	2
F	29	2	0	1	1
G	31.25	0	2	0	0
H	32	1	0	0	0
Pattern width		200	200	196.5	196

	Patterns			
Position	Pat_3	Pat_4	Pat_2	Pat_1
			widths	
1	25.75	25.75	25.75	18
2	25.75	25.75	25.75	23
3	16.5	16.5	17	23
4	16.5	16.5	17	23
5	16.5	16.5	17	23
6	16.5	16.5	17	29
7	16.5	16.5	18	29
8	16.5	16.5	31.25	32
9	29	29	31.25	
10	17	16.5		
Pattern width	196.5	196	200	200
Number of slitters moves	10	1	7	7

Fig. 2.7 Heuristic solution. Every step has a different color (top-down, left-to-right)

2.2.2   Conclusions

In its initial formulation, the hypothetical Slitter Moves Minimization Problem implies the search for optimal cutting patterns, their sequential arrangement, and the permutation of rolls within each pattern, all aimed at minimizing the overall count of slitter moves required.

This hypothetical SMMP poses a formidable challenge. Thus, it serves as an exemplary case where decomposition becomes necessary: resolving the CSP component first and subsequently addressing minimizing slitter moves. Nevertheless, a feedback loop from the slitter moves minimization back to the CSP is possible.

Even without its CSP aspect, the SMMP remains an intricate combinatorial problem. The absence of a polynomial-time algorithm for solving the SMMP further underscores its complexity. Remarkably, the SMMP appears even more intricate than the renowned Traveling Salesman Problem (TSP).

Given these challenges, employing a heuristic is the sole practical and efficient (albeit approximate) approach for solving the SMMP.

2.3   Skiving Stock Problem

That is how I discovered (or re-discovered) a Skiving Stock Problem (SSP). Along with cutting stock rolls into custom size finished rolls, there is a complementary process called *the skiving* process. The skiving process glues small stock rolls into larger finished rolls by widths (Fig. 2.8).

Indeed, the skiving process is opposite to the cutting process but serves a similar goal: to get the finals out of stock rolls most efficiently. The skiving process is not widely used in the paper industry though a few paper mills possess this technology. Also, SSP is a truly real-world problem though several researchers in Academia have expressed an interest in it.

In Johnson et al. (1997), we formulated the Skiving Stock Problem complementary to the CSP. Following that formulation, we define a skiving pattern.

Definition A *skiving pattern* is a combination of stock rolls skived so that the total width of those rolls exceeds the finished roll width on the amount sufficient to provide a necessary overlapping of every pair of adjacent stock rolls.

Fig. 2.8 Skiving process

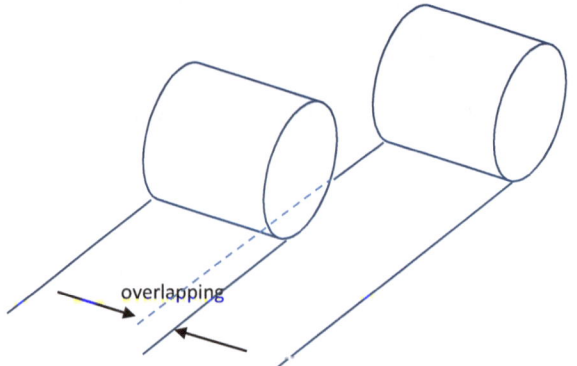

overlapping

$$\sum\nolimits_{i=1}^{m}\left(w_i^{stock}-e\right)x_i=w-e. \qquad (2.9)$$

Parameter $e \geq 0$ defines the overlapping width of two skived stock rolls. The overlapping width is any value in the range:

$$e^{\min} \leq e \leq e^{\max}.$$

To avoid the multiplication of two unknown variables, we use the following constraint instead of (2.9):

$$\sum\nolimits_{i=1}^{m}\left(w_i^{stock}-e^{\min}\right)x_i \geq w-e^{\min}. \qquad (2.10)$$

If a solution includes the constraint (2.10) as an inequality, then the final overlapping parameter is to be calculated:

$$e = e^{\min} + \delta / \left(\sum\nolimits_{i=1}^{m} x_i - 1\right)$$

where

$$\delta = \sum\nolimits_{i=1}^{m}\left(w_i^{stock}-e^{\min}\right)x_i - w + e^{\min}.$$

And we will get the equality (2.9). Of course, we must check:

$$e \leq e^{\max}.$$

Below (Fig. 2.9) is a schematic presentation of the two skived rolls:

Remark For simplicity of our consideration, we will drop the maximum overlapping parameter $e^{\max}$.

2.3.1   Covering Model

Sets:
$I = \{1, 2, \ldots, m\}$ is the set of stock rolls.
$K = \{1, 2, \ldots\}$ is the set of identical finished rolls.
Variables:
x_{ik} = number of stock rolls $i \in I$ covering a finished roll $k \in K$.
y_k = indication of covering finished roll $k \in K$.
The criterion is the number of finished rolls maximization:

Fig. 2.9 Skiving a finished
roll from two stock rolls

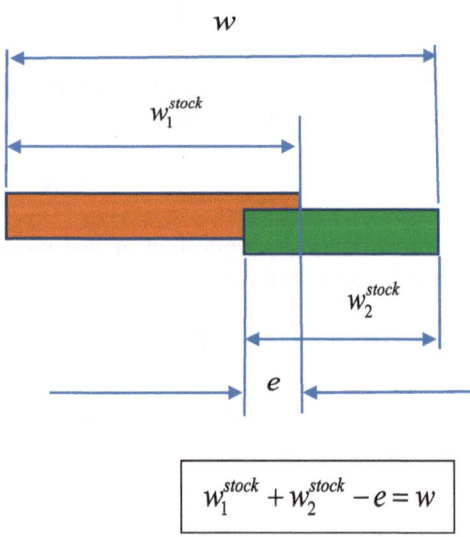

$$w_1^{stock} + w_2^{stock} - e = w$$

$$\textit{Maximize} \sum_{k \in K} y_k$$

The constraints are the following:
Number of small stock rolls covering a big finished roll:

$$\sum_{i \in I} \left(w_i^{stock} - e^{\min} \right) x_{ik} \ge \left(w - e^{\min} \right) y_k, \quad k \in K.$$

The inventory of small stock rolls is restricted:

$$\sum_{k \in K} x_{ik} \le b_i, \quad i \in I.$$

The numbers of small stock rolls are non-negative variables:

$$x_{ik} \in Z_+^1, \quad i \in I, k \in K.$$

The indications of covering big finished rolls are binaries:

$$y_k \in \{0, 1\}, \quad k \in K.$$

The covering model has the same flaw as a bin-packing model: a big
integrality gap.

The next step is toward a more practical model, a *pattern-based model*.

2.3.2 Pattern-Based Model

Given an inventory of small stock rolls, find skiving patterns and their activities to maximize the finished rolls' output.

$$Maximize \quad \mathbf{1}^T \mathbf{x}$$

$$Subject\ to \quad A\mathbf{x} \le \mathbf{b} \quad SPB(A, \mathbf{b})$$

$$\mathbf{x} \in Z_+^n$$

The LP relaxation and its dual counterpart are below:

Primal	Dual
Maximize $\mathbf{1}^T\mathbf{x}$	Minimize $\mathbf{b}^T\mathbf{u}$
Subject to $A\mathbf{x} \le \mathbf{b}$	Subject to $A^T\mathbf{u} \ge \mathbf{1}$
$\mathbf{x} \ge \mathbf{0}$	$\mathbf{u} \ge \mathbf{0}$

We suggest a dynamic column generation technique like that one for CSP. As in the case of CSP, the column selection step of the revised simplex algorithm is an overloaded function. Along with checking non-basic variables, we must consider a dynamic column generation.

A subproblem for the column generation is a classical covering problem:

$$Minimize \quad \mathbf{u}^T \mathbf{a}$$

$$Subject\ to \quad \left(\mathbf{w}^{stock} - \mathbf{1}e^{\min}\right)^T \mathbf{a} \ge w - e^{\min}$$

$$\mathbf{a} \in Z_+^m$$

Contrary to the knapsack problem, this covering problem is unbounded. However, we can impose simple bounds derived from the primal LP formulation:

$$\mathbf{a} \le \mathbf{b}.$$

The entire procedure looks like in Fig. 2.10.

An intriguing relationship exists between CSP and SSP. CSP, at first glance, resembles a covering problem as it aims to identify solutions that fulfill given demands. The subproblem of CSP takes the form of a knapsack problem, and its extension to multiple constraints resembles the nature of SSP.

Conversely, SSP exhibits similarities to a multi-constraint knapsack problem, as highlighted in reference (Zak and Dereksdottir 2001). The subproblem of SSP aligns with the structure of a classical covering problem, while its broader form involving multiple constraints bears resemblance to CSP (Fig. 2.11).

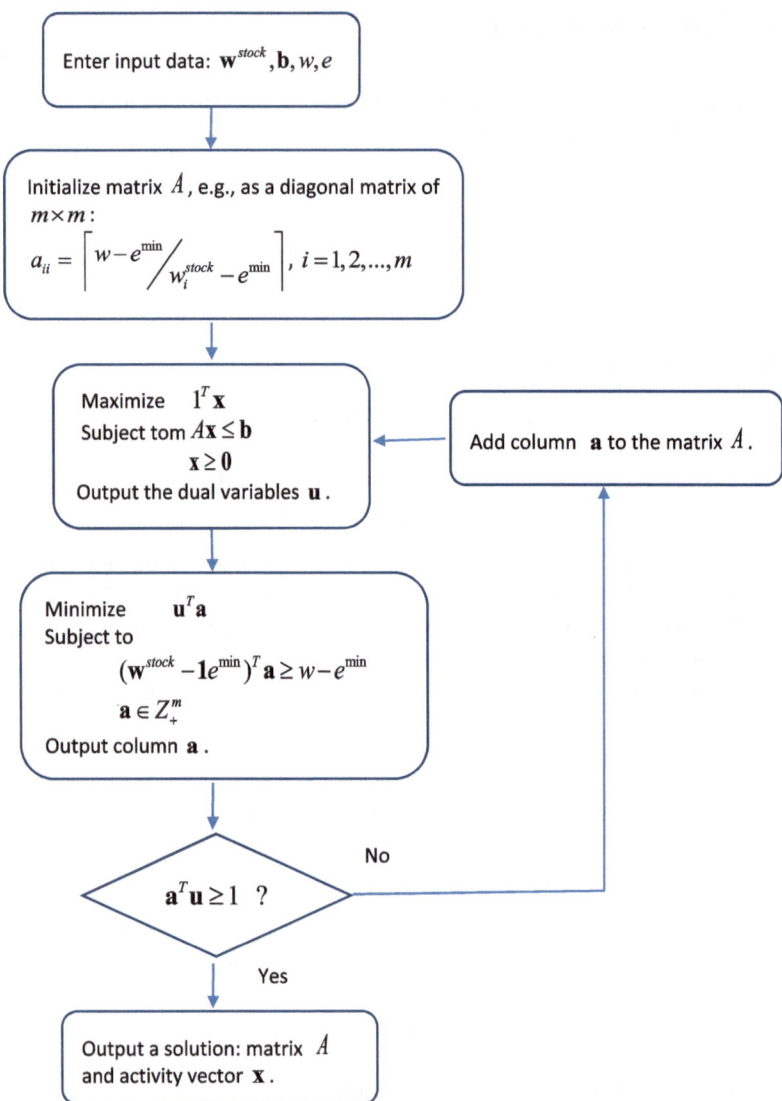

Fig. 2.10 Dynamic column generation for SSP

There is an exceptional case of SSP with zero overlapping $e^{\min} = 0$ (see Zak 2003). In this case SSP provides a lower bound of the corresponding CSP. In the perfect case of no covering loss a solution of SSP will be also a solution of CSP.

There is analogous to CSP, the *integer round-down property* for the SSP problem. The *integer round-down property* occurs if its optimal value is not below the greatest integer, less than or equal to its linear LP relaxation value:

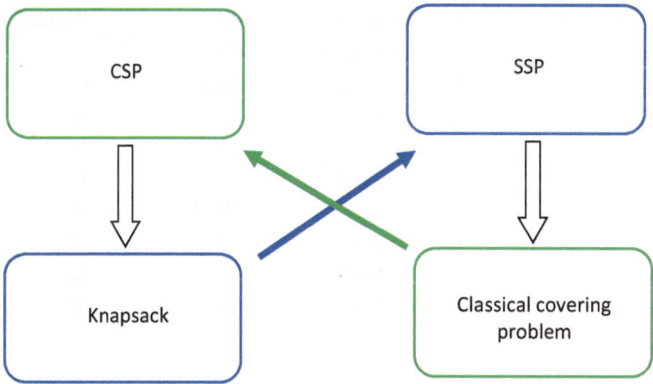

Fig. 2.11 Relationship between CSP and SSP

$$optimalValue \geq \lfloor LPoptimalValue \rfloor.$$

Most of the practical SSP problem falls into the valid round-down property class. There is another way to formulate this property.

For most practical SSPs, the integrality gap is less than 1.

$$| \, optimalValue - LPoptimalValue \, | \; \leq \; 1.$$

The author proves the round-down property holds for any SSP with two items (Zak 2003). Open conjecture: An integrality gap for any SSP is less than 2 (Zak 2003).

Example 2.7 We have constructed an SSP example that surpasses the complexity of a "typical" SSP scenario encountered in the paper industry. In the conventional paper industry SSP, the situation usually involves a maximum of two rolls being skived to form a finished roll. However, the subsequent example imposes no restrictions on the number of skived rolls allowed.

Furthermore, we present a parametric analysis of the outcomes derived from the overlapping value (parameter $e^{\min}$). In doing so, we unveil insights into how variations in this parameter influence the results.

In addition to this, we establish an intrinsic relation between SSP and CSP. To achieve this, we leverage the data introduced in Example 2.3. Nevertheless, the interpretation of this data will be slightly different. The following table (Table 2.15) shows the input data.

Furthermore, we vary overlapping parameter $e^{\min}$ from 0.0 to 3.0 with an increment of 0.5.

The plot below shows LP relaxation optimal values as a function of $e^{\min}$. It is not a straight line but a concave curve (Fig. 2.12).

The table below (Table 2.16) shows the exact numbers.

Below is an LP solution in the case $e^{\min} = 0$ (Table 2.17).

Table 2.15 SSP input data

i	Roll type	Width, w_i	Available amount, b_i
	Finished	200	–
1	Stock	16.5	32
2	Stock	17	47
3	Stock	18	26
4	Stock	23	64
5	Stock	25.75	32
6	Stock	29	37
7	Stock	31.25	22
8	Stock	32	15

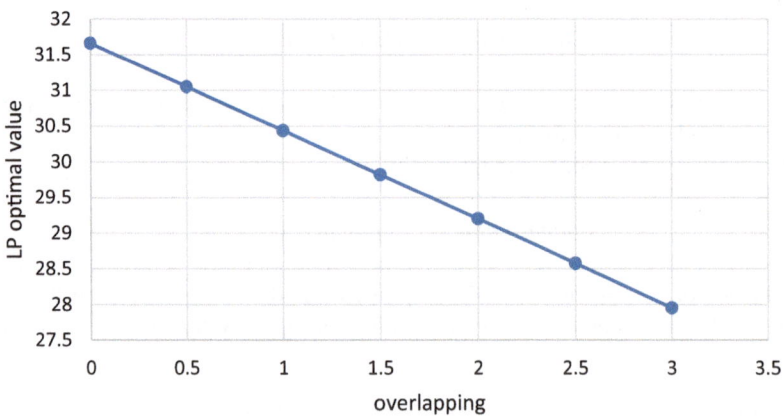

Fig. 2.12 LP optimal values as a function of $e^{\min}$

Table 2.16 LP relaxation optimal values as a function of $e^{\min}$

$e^{\min}$	LP optimal value
0	31.6575
0.5	31.0476
1	30.4347
1.5	29.8186
2	29.1995
2.5	28.5772
3	27.9518

We can observe that the objective function value of 31.6575 is the same as in the CSP case. Furthermore, the above SSP solution can also serve as a viable solution for the corresponding CSP. This fact highlights an intriguing "duality" between CSP and SSP!

Table 2.17 LP solution of the SSP in the case $e^{min} = 0$

		Patterns								Sch	Stock
		1	2	3	4	5	6	7	8	**31.6575**	
	Sets	3.074	1.156	5.333	5.094	2.704	5.5	1.242	7.554		
	Sock widths									Sch	Stock
1	16.5	0	10	1	0	0	0	0	2	32	32
2	17	0	1	0	9	0	0	0	0	47	47
3	18	4	1	0	1	0	0	6	0	26	26
4	23	0	0	0	0	1	2	4	6	64	64
5	25.75	0	0	6	0	0	0	0	0	32	32
6	29	0	0	1	1	5	1	0	1	37	37
7	31.25	0	0	0	0	0	4	0	0	22	22
8	32	4	0	0	0	1	0	0	0	15	15
	Pattern width	200	200	200	200	200	200	200	200		
	Overlapping width	0	0	0	0	0	0	0	0		
	Finished roll width	200	200	200	200	200	200	200	200		

I present another LP solution in the case $e^{min} = 1$ (Table 2.18).

First, we can observe that the optimal value of 30.4347 is lower than in the previous case (31.6575). Second, it is evident that each pattern width accommodates an additional width, known as overlapping width, corresponding to the number of skived rolls minus one. For instance, the first pattern skives seven rolls, requiring $6 \times 1" = 6"$ of overlapping width for its six pairs. Similarly, the second pattern skives eight rolls, necessitating $7 \times 1" = 7"$ of overlapping width for its seven pairs, and so on.

Rounding this LP solution results in an objective value of 29 and 13 unused stock rolls. Solving another skiving problem with the remaining stock rolls yields an additional pattern (pattern #9 below) and improves the solution to 30 with only four unused stock rolls (Table 2.19).

Here is a heuristic solution for the same case $e^{min} = 1$ (Table 2.20).

We can observe that the heuristic solution exhibits fewer patterns and utilizes fewer stock rolls but incurs higher trim loss. Additionally, it is worth noting that SSP possesses a round-down property.

As in the case of CSP, we introduce the following metrics for SSP (Table 2.21).

Remarks

- Please note that SSP has no entity such as "ordered amount" and related to it "overrun" and "underrun."
- Trim loss includes overlapping loss.
- The following relations between metrics are valid:

$$Input = Output + TrimLoss$$

$$Efficiency = \frac{Output}{Input} \, 100\%.$$

Cutting and skiving processes occur in certain practical scenarios in two distinct stages. In the first stage, large rolls are cut into finished and intermediate rolls. In the second stage, these intermediate rolls are then skived to produce finished rolls.

2.3.3 Conclusions

The Skiving Stock Problem (SSP) revolves around optimizing finished roll output by identifying suitable skiving patterns. SSP represents a relatively recent addition to the realm of Operations Research challenges. Notably, SSP complements the Cutting Stock Problem (CSP).

Consequently, integrating SSP into the same software package as CSP appears logical. The key distinction lies in the auxiliary problems: CSP involves a knapsack

Table 2.18 LP solution of the SSP in the case $e^{min} = 1$

		Patterns									Stock
		1	2	3	4	5	6	7	8	**30.4347**	
	Sets	3.963	0.89	3.983	3.2	7.532	5.688	1.1	4.078	Sch	Stock
	Stock widths										
1	16.5	0	0	0	10	0	0	0	0	32	32
2	17	0	1	10	0	0	0	2	1	47	47
3	18	0	0	1	0	1	1	8	0	26	26
4	23	0	1	1	2	7	0	0	0	64	64
5	25.75	0	4	0	0	0	5	0	0	32	32
6	29	6	0	0	0	1	1	0	0	37	37
7	31.25	0	0	0	0	0	1	0	4	22	22
8	32	1	2	0	0	0	0	1	2	15	15
	Pattern width	206	207	211	211	208	207	210	206		
	Overlapping width	6	7	11	11	8	7	10	6		
	Finished roll width	200	200	200	200	200	200	200	200		

Table 2.19 LP-based solution of the SSP in the case $e^{min} = 1$

		Patterns										
		1	2	3	4	5	6	7	8	9		
	Sets	4	1	4	3	7	5	1	4	1	**30**	
	Stock Widths										Sch	Stock
1	16.5	0	0	0	10	0	0	0	0	2	32	32
2	17	0	1	10	0	0	0	2	1	0	47	47
3	18	0	0	1	0	1	1	8	0	0	24	26
4	23	0	1	1	2	7	0	0	0	3	63	64
5	25.75	0	4	0	0	0	5	0	0	3	32	32
6	29	6	0	0	0	1	1	0	0	1	37	37
7	31.25	0	0	0	0	0	1	0	4	0	21	22
8	32	1	2	0	0	0	0	1	2	0	15	15
	Pattern width	206	207	211	211	208	207	210	206	208		

Table 2.20 Heuristic solution of the SSP in the case $e^{min} = 1$

		Patterns						
		1	2	3	4	5		
	Sets	18	8	2	1	1	**30**	
	Stock Widths						Sch	Stock
1	16.5	1	1	3	0	0	32	32
2	17	2	1	0	1	2	47	47
3	18	0	1	0	6	10	24	26
4	23	3	1	0	1	0	63	64
5	25.75	0	4	0	0	0	32	32
6	29	2	0	0	1	0	37	37
7	31.25	1	0	2	0	0	22	22
8	32	0	1	3	1	0	15	15
	Pattern width	208.75	209.5	208	209	214		

Table 2.21 Metrics definition along with specific values for Example 2.7 and $e^{min} = 1$

Metric	Definition	LP solution	IP solution	Heuristic
Input	$(\mathbf{w}^{stock})^T A\mathbf{x}/w$	31.6575	31.206	31.3625
Output	$\mathbf{1}^T \mathbf{x}$	30.4347	30	30
UpperBound	$(\mathbf{w}^{stock} - \mathbf{1}e^{min})^T \mathbf{b}/(w - e^{min})$	30.4347	30.4347	30.4347
TrimLoss	$[(\mathbf{w}^{stock})^T A/w - \mathbf{1}^T]\mathbf{x}$	1.2228	1.206	1.3625
Efficiency	$w\mathbf{1}^T \mathbf{x}/(\mathbf{w}^{stock})^T A\mathbf{x}$	96.137%	96.134%	95.656%
NumPatterns	$\mathbf{1}^T \text{ sgn}(\mathbf{x})$	8	9	5

problem, whereas SSP entails a more classical covering problem. Remarkably, solving the SSP problem with $e^{min} = 0$ could yield a solution for the CSP problem.

2.4   Two-Stage Cutting Stock Problem

A two-stage cutting stock problem is a generalization of a single-stage CSP, introducing an additional layer of complexity when the cutting process is distributed across two stages. In the first stage, stock rolls are cut into intermediate sizes whose dimensions remain unknown. The first stage output may include finished rolls as well. The second stage involves further cutting these intermediate rolls to produce finished rolls.

The two-stage CSP is a real-world problem lacking a traditional counterpart in textbooks. There are some research papers, e.g., by J.M. Valerio de Carvalho (De Carvalho and Rodrigues 1995). Why does a single-stage cutting process cannot simply replace a two-stage approach? The answer lies in the technical and technological constraints of the cutting machines. For instance, there may not be enough slitters available in the machines (winders) to cut small-width rolls in a single pass. Alternatively, producing coated paper might be more efficient when utilizing wider rolls.

In our discussion, we will focus on a two-stage CSP scenario involving a single type of stock roll. In the first stage, Machine 1 is responsible for cutting these stock rolls into multiple intermediate rolls (as illustrated in Fig. 2.13). In the second stage, Machine 2 performs the final cutting to create finished rolls. Every intermediate roll must contain a minimum edge requirement to make the second stage of the cutting process feasible.

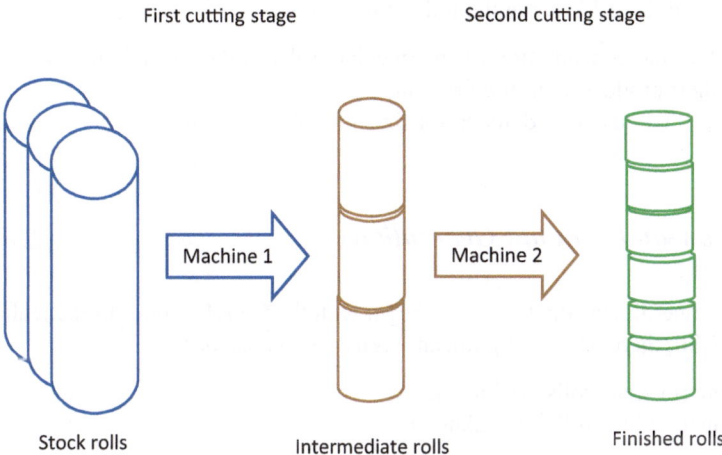

Fig. 2.13 Two-stage cutting process

2.4.1   Mathematical Model

We will reference the model described in Zak (2002a, b).

$$Minimize \quad \begin{pmatrix} 1^T & 0^T \end{pmatrix} \begin{pmatrix} \mathbf{x}^1 \\ \mathbf{x}^2 \end{pmatrix} \tag{2.11}$$

$$Subject\ to \quad \begin{pmatrix} A_{11} & A_{12} \\ 0 & A_{22} \end{pmatrix} \begin{pmatrix} \mathbf{x}^1 \\ \mathbf{x}^2 \end{pmatrix} \geq \begin{pmatrix} \mathbf{0} \\ \mathbf{b} \end{pmatrix} \tag{2.12}$$

$$\mathbf{x}^1 \in Z_+^{n_1}; \mathbf{x}^2 \in Z_+^{n_2} \tag{2.13}$$

where A_{11} and A_{22} are matrices of cutting patterns for the first and second stages, respectively. The size of A_{11} is $l \times n_1$ and size of A_{22} is $m \times n_2$, where l and m are the numbers of intermediate and finished rolls, respectively; n_1 and n_2 are the numbers of the first and the second stage cutting patterns, respectively.

A_{12} is a matrix where every column j is a vector $(0, \ldots, 0, -1, 0, \ldots, 0)^T$, where exactly one non-zero element "-1" appears in the position i corresponding to the intermediate roll i that should be cut according to the cutting pattern defined by column j in matrix A_{22}. The size of matrix A_{12} is $l \times n_2$.

$\mathbf{x}^1$ and $\mathbf{x}^2$ are vectors of the first-stage and second-stage pattern activities, respectively.

The objective function (2.11) aims to minimize the number of stock rolls, as defined by the respective term $1^T \mathbf{x}^1$.

The constraints (2.12) serve the following purposes:

- Ensure that the consumption of intermediate rolls in the second stage does not exceed their production in the first stage.
- Guarantee that customer demand for finished rolls is satisfied.

2.4.2   Row-and-Column Generation

Just as with one-dimensional CSP, we begin with the LP relaxation. In the realm of two-stage CSP, there are two significant scenarios to consider:

1. When intermediate rolls are known.
2. When intermediate rolls are unknown.

For situations involving known intermediate rolls, we employ the column generation technique while bearing in mind that we have columns (patterns) of two distinct types corresponding to the two stages of the cutting process.

Now, let us focus on the scenario of unknown intermediate rolls. In this case, the dual problem associated with the LP relaxation of the primal problem (2.11–2.13) is as follows:

$$\text{Maximize} \quad \begin{pmatrix} 0^T & b^T \end{pmatrix} \begin{pmatrix} \mathbf{u}^1 \\ \mathbf{u}^2 \end{pmatrix} \tag{2.14}$$

$$\text{Subject to} \quad \begin{pmatrix} A_{11}^T & 0 \\ A_{12}^T & A_{22}^T \end{pmatrix} \begin{pmatrix} \mathbf{u}^1 \\ \mathbf{u}^2 \end{pmatrix} \leq \begin{pmatrix} 1 \\ 0 \end{pmatrix} \tag{2.15}$$

$$\begin{pmatrix} \mathbf{u}^1, \mathbf{u}^2 \end{pmatrix}^T \geq 0 \tag{2.16}$$

Here, vectors $\mathbf{u}^1$ and $\mathbf{u}^2$ are vectors of dual variables corresponding to intermediate and finished rolls, respectively.

The dual problem (2.14–2.16) leads to three auxiliary problems we should solve at the column selection step of the revised simplex algorithm.

The first type of auxiliary problem generates a cutting pattern for the first stage of the cutting process. The auxiliary problem is the same knapsack problem as we have in a single-stage CSP:

$$\text{Maximize} \quad \mathbf{u}^1 \mathbf{a}^{11}$$

$$\text{Subject to} \quad \mathbf{w}^{inter} \mathbf{a}^{11} \leq w^{stock}, \quad KS\big(\mathbf{u}^1, \mathbf{w}^{inter}, w^{stock}\big)$$

$$\mathbf{a}^{11} \in Z_+^l.$$

Knapsack $KS(\mathbf{u}^1, \mathbf{w}^{inter}, w^{stock})$ generates a column $\mathbf{a}^{11}$ for the first stage.

The second type of auxiliary problem generates a cutting pattern for the second stage of the cutting process. Please note that this knapsack works with every intermediate roll w_j^{inter}, $j = 1, 2, \ldots, l$.

$$\text{Maximize} \quad \mathbf{u}^2 \mathbf{a}^{22}$$

$$\text{Subject to} \quad \mathbf{w} \mathbf{a}^{22} \leq w_j^{inter} - e^{min}, \quad KS\big(\mathbf{u}^2, \mathbf{w}, w_j^{inter} - e^{min}\big)$$

$$\mathbf{a}^{22} \in Z_+^m$$

where e^{min} is the minimum edge required for an intermediate roll to be cut at the second stage.

Knapsack $KS\big(\mathbf{u}^2, \mathbf{w}, w_j^{inter} - e^{min}\big)$ generates a column $\mathbf{a}^{22}$ for the second stage.

Beyond these two, we face a third auxiliary problem associated with the simultaneous generation of new intermediate rolls and cutting patterns.

Assumption. At most, the row-and-column generation algorithm generates one new intermediate roll at a time.

In theory, this assumption limits our options for generating intermediate rolls. In fact, there may be instances where no new cutting patterns can be created using only a single new intermediate roll. In the first stage of a cutting pattern, it is necessary to incorporate at least two new intermediate rolls. Nevertheless, we justify this

assumption based on extensive observations across a broad spectrum of real-world
two-stage CSPs.

$$Maximize \quad \mathbf{u}^1\mathbf{a}^{11} + ua \tag{2.17}$$

$$Subject\ to \quad \mathbf{w}^{inter}\mathbf{a}^{11} + wa \leq w^{stock} \tag{2.18}$$

$$\mathbf{wa}^{22} \leq w - e^{\min} \tag{2.19}$$

$$\mathbf{u}^2\mathbf{a}^{22} = u \tag{2.20}$$

$$\mathbf{a}^{11} \in Z_+^m;\ \mathbf{a}^{22} \in Z_+^m \tag{2.21}$$

$$w \in \left[w_{\min}^{inter}, w_{\max}^{inter}\right] \tag{2.22}$$

$$u \geq 0 \tag{2.23}$$

$$a \in Z_+^1 \tag{2.24}$$

where
$\left[w_{\min}^{inter}, w_{\max}^{inter}\right]$ is a given range for the new intermediate roll width,
w is a variable corresponding to the width of the new intermediate roll,
u is a dual variable corresponding to the new intermediate roll,
a is a variable corresponding to the number of the new intermediate roll occurrences in a first stage pattern.

The objective function (2.17) maximizes the intermediate rolls price.
Constraint (2.18) restricts the first stage pattern width.
Constraint (2.19) restricts the second stage pattern width.
Constraint (2.20) connects dual variables.
Constraints (2.21–2.24) define the variables regions.

There are two nonlinearities in the model (2.17–2.24): ua and wa. Parametrization of integer variable a copes with these nonlinearities. The finite set of a is $\{1, 2, \ldots, \lfloor \left(w^{stock} - e^{\min}\right)/w_{\min}^{inter} \rfloor \}$. So, we must solve the model (2.17–2.24) for every value of a.

2.4.3   A Modified Column Selection in the Revised Simplex Algorithm

Step 1. Solve Knapsack 1. If the optimal value of the functional exceeds 1.0, solution $\mathbf{a}^{11}$ is a column to enter the basis at the pivoting step of the revised simplex algorithm. Otherwise, go to the next step.

Step 2. Solve Knapsack 2 for each existing intermediate roll w_j^{inter}. If the optimal value of the functional exceeds 1.0, solution $\mathbf{a}^{22}$ is a column to enter the basis at the pivoting step of the revised simplex algorithm. Otherwise, go to the next step.

Step 3. Solve Knapsack 3. If the optimal value of the functional exceeds 1.0, we should:

- Expand the basis matrix B by one row and one column: $\begin{pmatrix} B & \mathbf{a}^{11} \\ \mathbf{0}^T & -1 \end{pmatrix}$.

- Expand the inverse matrix B^{-1} accordingly: $\begin{pmatrix} B^{-1} & B^{-1}\mathbf{a}^{11} \\ \mathbf{0}^T & -1 \end{pmatrix}$.

- Pick the other column $\mathbf{a}^{22}$ to enter the basis at the pivoting step of the revised simplex algorithm.

If the optimal value of the functional does not exceed 1.0, and the reduced costs of slack variables are non-negative, then the current solution is optimal.

You must develop a slightly more sophisticated (than in the case of a single-stage CSP) rounding algorithm.

2.4.4   Heuristic Algorithm

The heuristic concept involves transforming a two-stage problem into a single stage. In this scenario, we assume that a single cutting stage produces finished rolls directly from stock rolls. The primary requirement is to allocate additional space for the edges of the intermediate rolls. We must understand how many intermediate rolls are in a pattern to do this. The minimal number is the following:

$$n = \lceil w^{stock} / w^{inter}_{max} \rceil \tag{2.25}$$

So, the effective stock roll width must be:

$$w^{stock}_{eff} = w^{stock} - ne^{min}. \tag{2.26}$$

An equivalent single-stage model is:

$$Minimize \ \sum_{j \in J} x_j \tag{2.27}$$

$$Subject \ to \ \sum_{j \in J} \mathbf{a}_j x_j > \mathbf{b} \tag{2.28}$$

$$x_j \in Z^1_+ \tag{2.29}$$

where
$\mathbf{a}_j$ is a composite pattern of finished rolls over effective stock size w^{stock}_{eff}.
$\mathbf{b}$ is finished rolls demand.
$\mathbf{x}$ is a vector of pattern activities.

Now we must check validity of every pattern $\mathbf{a}_j, j \in J$ in terms of its transforming it into a pattern of intermediate rolls. For this purpose, we formulate an auxiliary CSP problem:

$$\text{Minimize} \quad \mathbf{1}^T \mathbf{x}^j \tag{2.30}$$

$$\text{Subject to} \quad A^j \mathbf{x}^j = \mathbf{a}_j \tag{2.31}$$

$$\mathbf{x}^j \in Z_+^m \tag{2.32}$$

where
A^j is a matrix of feasible patterns of finished rolls over the knapsack size $w_{\text{max}}^{inter} - e^{\text{min}}$.
$\mathbf{x}^j$ is a vector of pattern activities.

Remark Please note that the constraints are in the equality form because the exact number of finished rolls must be consumed.

Example 2.8 Two-stage CSP. We continue exploring the same data set. Let us assume intermediate rolls fall within the range of [55.00, 75.00]. Each intermediate roll must have an obligatory edge measuring 1.00.

The target number of intermediate rolls in a pattern according to the formula (2.25) is:

$$n = \left\lceil w^{stock} / w_{\text{max}}^{inter} \right\rceil = \lceil 200/75 \rceil = 3.$$

According to formula (2.26) we should reserve $3 \times 1 = 3$ edge for intermediate rolls. So, the effective stock width is $200 - 3 = 197$.
Take a composite pattern generated by the equivalent single stage CSP:

$$\mathbf{a}_j = (0, 2, 2, 3, 0, 2, 0, 0)^T.$$

The pattern width is 197. Now we solve model (2.30–2.32) with the knapsack $KS\left(\mathbf{w}, \mathbf{a}_j, w_{\text{max}}^{inter} - e^{\text{min}}\right)$.
The solution is in Fig. 2.14.
Highlighted in blue is the matrix A^j and highlighted in green is a given vector $\mathbf{a}_j$.
So, we have generated two intermediate rolls of 65 as a composition of 17, 18, and 29 widths finished rolls plus one-unit measure for edge, and on intermediate roll of 70 as a composition of 23 width finished roll by three plus one-unit measure for edge. Their width match perfectly the stock size: $2 \times 65 + 1 \times 70 = 200$.

Example 2.9 Two-stage CSP. In this setting, it is mandated that all finished rolls must be derived from the second stage. The intermediate rolls are the products of the first stage. Furthermore, it is a requirement that all intermediate rolls produced in the first stage must be utilized in the second stage.

Fig. 2.14 Solution of the knapsack $KS\left(\mathbf{w}, \mathbf{a}_j, w_{max}^{inter} - e^{min}\right)$

		Patterns			
		1	2		
Item	Sets	2	1		
	Widths			Sch	Ord
A	16.5				
B	17	1		2	2
C	18	1		2	2
D	23		3	3	3
E	25.75				
F	29	1		2	2
G	31.25				
H	32				
Pattern width		64	69		
Intermediate rolls		65	70		

The heuristic has yielded a solution (Fig. 2.15) that generates four first-stage patterns, 11 second-stage patterns, and seven intermediate rolls (I1 through I7).

In this figure, matrices A_{11} and A_{22} are highlighted in blue, while matrix A_{12} is marked in green. As we can see (column "Sch"), the intermediate rolls generated in the first stage are fully utilized in the second stage. There is a slight surplus in the finished rolls. Both stages contribute to trim loss, with the second stage's trim loss being obligatory edge loss.

The table below (Table 2.22) lists frequently used criteria and their values for the heuristic solution.

The following relations between metrics are valid:

$$Input = Output + TrimLoss$$

$$Output = OrdQty + OverRun - UnderRun$$

$$Efficiency = \frac{Output}{Input} 100\%.$$

2.4.5   Conclusions

A two-stage cutting process adds a layer of complexity and challenge to the realm of CSP. This intricate issue requires the discovery of cutting patterns in both stages and the determination of intermediate roll widths.

Several factors contribute to facilitating the multistage CSP, including:

Item	Widths	Patterns															Sch	Ord
	Stage #	1	1	1	1	2	2	2	2	2	2	2	2	2	2	2		
	Sets	14	7	4	8	5	5	11	9	23	5	3	18	4	14	2		
		$\mathbf{x}^1$				$\mathbf{x}^2$												
	200	-1	-1	-1	-1												-33	
I1	60.75			1									-1				0	
I2	63.5				2		-1		-1								0	
I3	65	2				-1			-1								0	
I4	66.25		2	1									-1				0	
I5	67.5		1			-1									-1		0	
I6	70	1												-1			0	
I7	72.75			1	1			-1			-1						0	
A	16.5					1				1		1				3	34	32
B	17						1	1	1				2			1	47	47
C	18					1			1								28	26
D	23									2	1		3				66	64
E	25.75						1			1	1	1					34	32
F	29						1	1	1								37	37
G	31.25						2										22	22
H	32				2	1											15	15
	Trim Loss(-)	0	0	-0.25	-0.25	-1	-1	-1	-1	-1	-1	-1	-1	-1	-1	-1		

Fig. 2.15 Heuristic solution of the two-stage CSP

Table 2.22 Metrics definition along with specific values for Example 2.8

Metric	Definition	Heuristic solution
Input	$\mathbf{1}^T\mathbf{x}^1$	33
Output	$\mathbf{w}^T A_{22}\mathbf{x}^2/w^{stock}$	32.49
OrdQty	$\mathbf{w}^T\mathbf{b}/w^{stock}$	31.6575
LowerBound	$\mathbf{w}^T\mathbf{b}/w^{stock}$	31.6575
TrimLoss at 1st stage	$(\mathbf{1}^T - \mathbf{w}^{interT}A_{11}/w^{stock})\mathbf{x}^1$	0.015
TrimLoss at 2nd stage	$(\mathbf{w}^{interT} - \mathbf{w}^T A_{22})/w^{stock})\mathbf{x}^2$	0.495
TrimLoss	$(\mathbf{1}^T - \mathbf{w}^{interT}A_{11}/w^{stock})\mathbf{x}^1 + (\mathbf{w}^{interT} - \mathbf{w}^T A_{22})/w^{stock})\mathbf{x}^2$	0.51
OverRun	$\mathbf{w}^T \max(\mathbf{0}, A_{22}\mathbf{x}^2 - \mathbf{b})/w^{stock}$	0.8325
UnderRun	$\mathbf{w}^T \max(\mathbf{0}, \mathbf{b} - A_{22}\mathbf{x}^2)/w^{stock}$	0.0
Efficiency, %	$w^{stock}\mathbf{1}^T\mathbf{x}^1/\mathbf{w}^T A_{22}\mathbf{x}^2$	98.455
NumPatterns	$\mathbf{1}_1^T \operatorname{sgn}(\mathbf{x}^1) + \mathbf{1}_2^T \operatorname{sgn}(\mathbf{x}^2)$	15

- The list of potential intermediate sizes is established and finite.
- Although the intermediate sizes remain unknown, each intermediate roll comprises identical finished rolls.

The existing algorithms, particularly an elegant row-and-column generation method, bode well for the efficient resolution of practical two-stage models.

Empirical trials (Zak 2002a, b) vividly showcase the procedure's effectiveness and its solutions' exceptional caliber.

While our demonstration has spotlighted the technique within the context of a two-stage CSP, it is imperative to note that the same technique is equally applicable to the broader domain of a multistage CSP.

2.5   Warehouse Storage Space Problem

The Warehouse Storage Space Problem (WSSP) addresses the efficient utilization of storage space within a multi-item setting of a warehouse. The warehouse's inventory encompasses various items, all integral to order fulfillment. The core challenge entails strategically coordinating the arrival of each vendor-supplied item at the warehouse. The objectives are to minimize storage capacity requirements while meeting customer demand. Notably, the author of this study is the pioneer in formulating the WSSP, blurring any distinction between "theoretical" and "practical" formulations.

This study introduces two distinct models to tackle the WSSP. The initial model employs a continuous approach rooted in a captivating mathematical interpretation. Specifically, it seeks to minimize the aggregate sum of inverse sawtooth functions. The subsequent model adopts a discrete mathematical programming approach.

2.5.1   Continuous Warehouse Storage Space Model

I presented this model at the 23rd International Symposium on Mathematical programming (Zak 2018).
Let I be a set of items stored in a warehouse.
We assume the following information is available for every type of stored item:

- The maximum number of items $x_i^{\max}$, $i \in I$, arrived to the storage with every shipment from a vendor to the warehouse.
- The safety stock $q_i^{\min}$, $i \in I$.
- The items receiving rate r_i [items/time], $i \in I$.

The item replenishment cycle h_i is the following:

$$h_i = \frac{x_i^{\max}}{r_i}. \tag{2.33}$$

The item inventory dynamics is an inverse sawtooth function with period h_i (Fig. 2.16).
The analytical formula for this inverse sawtooth function is:

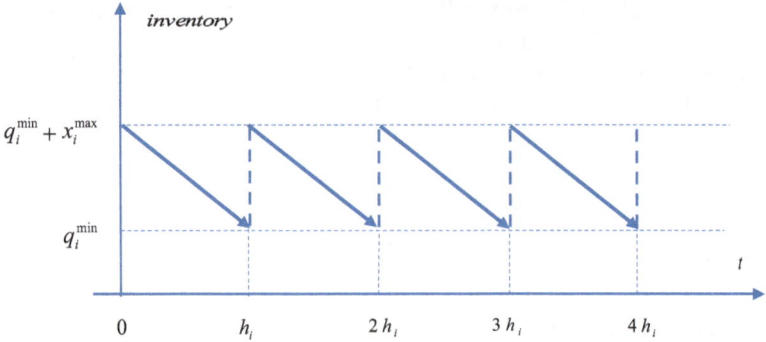

Fig. 2.16 An inverse sawtooth function

$$q_i(t) = q_i^{\min} + x_i^{\max} - r_i t + x_i^{\max} \cdot \left\lfloor \frac{t}{h_i} \right\rfloor, \tag{2.34}$$

where

$q_i(t)$ is an i-item inventory function of time $-\infty < t < +\infty$, $i \in I$.

$\lfloor x \rfloor$ is the floor function: the greatest integer that is less than or equal to x.

Equation (2.34) assumes that the peaks coincide with time points $t = h_i k$, $k = 0, \pm 1, \pm 2, \ldots$. So, the maximum storage capacity for a single item is:

$$q_i^{\min} + x_i^{\max}.$$

Due to the function uniqueness, the safety stock $q_i^{\min}$ never reaches, i.e., $q_i(t) > q_i^{\min}$, $\forall t$.

To make the formula (2.34) more generic, we introduce a phase shift $0 \le \varphi_i < h_i$ and replace parameter r_i according to (2.33) as $x_i^{\max}/h_i$.

$$q_i(t) = q_i^{\min} + x_i^{\max}\left(1 - \frac{t - \varphi_i}{h_i} + \left\lfloor \frac{t - \varphi_i}{h_i} \right\rfloor\right). \tag{2.35}$$

Time points $t = h_i k + \varphi_i$, $k = 0, \pm 1, \pm 2, \ldots$ correspond to the peaks (Fig. 2.17).

Since every item has its inverse sawtooth function, the problem is how to spread those functions in time to minimize peaks of the total inventory.

Example 2.10 WSSP. Let us assume that two items from two vendors comprise the inventory. All parameters $\{q^{\min}, x^{\max}, h\}$ but phase shift φ of these two items are identical. The question is how to choose the phase shifts φ_1 and φ_2 to minimize the storage capacity requirement. In other words, how to spread out the inverse sawtooth functions along the time axis to provide the lowest maximum sum of those functions.

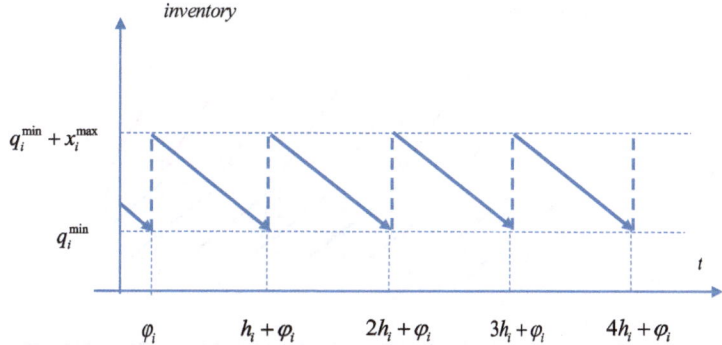

Fig. 2.17 An inverse sawtooth function with a phase shift

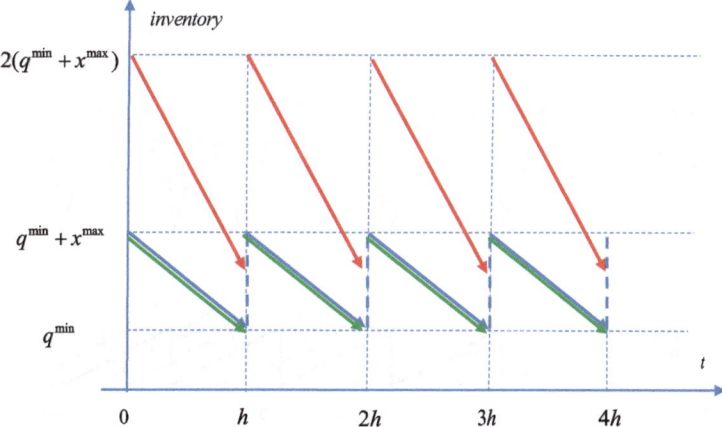

Fig. 2.18 Sum of two identical inverse sawtooth functions with identical phase shifts

The graph in Fig. 2.18 demonstrates the worst-case scenario solution when $\varphi_1 = \varphi_2$. Thus, the capacity requirement is $2(q^{\min} + x^{\max})$.

The graph in Fig. 2.19 illustrates the optimal solution $\varphi_1^{opt} = 0$ and $\varphi_2^{opt} = h/2$.

The sum (in red) of two inverse sawtooth functions (in blue and green correspondingly) oscillates between $2q^{\min} + 0.5x^{\max}$ and $2q^{\min} + 1.5x^{\max}$. Thus, capacity requirement is $2q^{\min} + 1.5x^{\max}$.

Remarks

- Here, we assume that only the relative positioning of the curves is essential; that is why in all cases, we can choose one item with a zero-phase shift, normally the first one $\varphi_1 = 0$.
- The optimal sum is a periodic function with the least period $h/2$. Generally, the sum might be periodic if a common multiplier of $\{h_i\}$ exists (Radcliffe 2013). The common multiplier will be a period but not necessarily the least one.

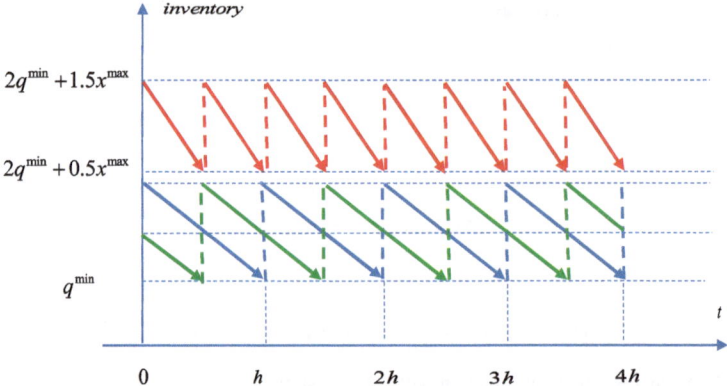

Fig. 2.19 Sum of two identical inverse sawtooth functions with optimal phase shifts

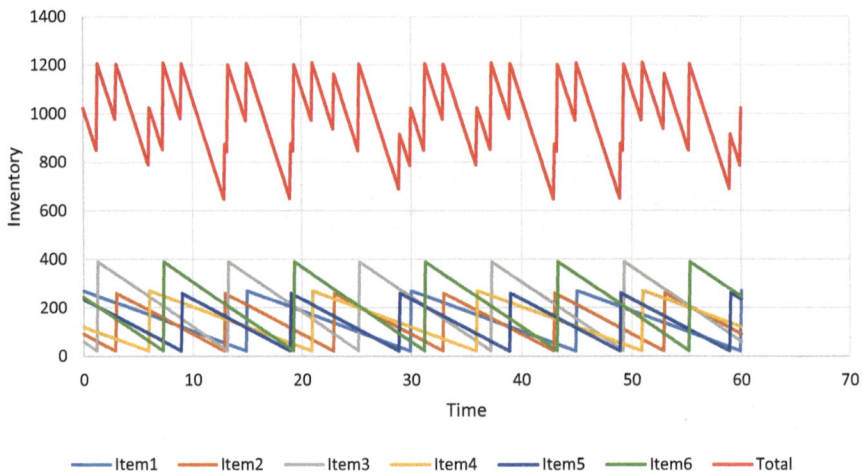

Fig. 2.20 A six-item case of the inventory dynamics

In a more general case, a sum of the inverse sawtooth functions may look like in Fig. 2.20.

Mathematical Formulation

Introduce:

$I = \{1, 2, \ldots, m\}$ is the set of items in the storage inventory.
$\mathbf{h} = (h_1, h_2, \ldots, h_m)^T$ is the vector of replenishment cycles for all items.
$\boldsymbol{\varphi} = (\varphi_1, \varphi_2, \ldots, \varphi_m)^T$ is the vector of item phase shifts.
w_i is the volume of a unit of item $i \subset I$.

$t^{\max}$ is the planning horizon which normally includes several periods, e.g., $t^{\max} = N\max_i\{h_i\}$ where N is the number of replenishment cycles corresponding to the maximum period.

In general, the model is about finding the following mini-max:

$$\min_{0 \le \varphi < h} \max_{0 \le t \le t^{\max}} \sum_{i \in I} w_i \left(q_i^{\min} + x_i^{\max} \left(1 - \frac{t - \varphi_i}{h_i} + \left\lfloor \frac{t - \varphi_i}{h_i} \right\rfloor \right) \right) \quad (2.36)$$

There are two groups of continuous variables:

- A single continuous variable—time t,
- A vector of continuous variables—phase shifts $\boldsymbol{\varphi} = (\varphi_1, \varphi_2, \ldots, \varphi_m)^T$.

The model (2.36) is a mini-max model with a non-linearity caused by the floor function $\lfloor x \rfloor$. An analytical solution to this problem is unknown.

For further discussion, we formulate and prove theorems.

Let us consider the inner problem in (2.36):

$$\max_{0 \le t \le t^{\max}} \sum_{i \in I} w_i \left(q_i^{\min} + x_i^{\max} \left(1 - \frac{t - \varphi_i}{h_i} + \left\lfloor \frac{t - \varphi_i}{h_i} \right\rfloor \right) \right) \quad (2.37)$$

Theorem 2.3 Given vector $\boldsymbol{\varphi} = (\varphi_1, \varphi_2, \ldots, \varphi_m)^T$, the optimal solution of the inner problem (2.37) belongs to the finite discrete set of the following time points:

$$\cup_{i \in I} \left\{ h_i k + \varphi_i, \quad k = 0, 1, 2, \ldots, \left\lfloor \frac{t^{\max}}{h_i} \right\rfloor \right\}.$$

The theorem is intuitive since every peak of the sum of the inverse sawtooth functions corresponds to a peak of one of its terms. Appendix B presents the formal proof.

Theorem 2.4 In the case of m items with identical parameters $\{q^{\min}, x^{\max}, h, w\}$, an optimal solution to the original problem (2.36) is the following:

$$\varphi_i^{opt} = \frac{h}{m}(i - 1), \quad i = 1, 2, \ldots, m,$$

and the optimal value of the objective function is the following:

$$w \left(mq^{\min} + \frac{m + 1}{2} x^{\max} \right).$$

Appendix B presents the formal proof.

In the worst-case scenario when all items arrive at the same time $\varphi_1 = \varphi_2 = \cdots = \varphi_m = 0$ the value of the objective function is the following:

$$wm\left(q^{\min} + x^{\max}\right).$$

So, the gain of optimality $w\frac{m-1}{2}x^{\max}$ is significant and grows linearly with the number of items.

Corollary. In the optimal solution to the original problem (2.36) having identical parameters $\{q^{\min}, x^{\max}, h, w\}$, the sum is a periodic function with the shortest period h/m.

We propose an MIP-based approach to solving the problem (2.36). We discretize time and phase shifts to convert the original continuous model (2.36) to the MIP model.

Time Discretization

Designate Δt as an elementary time interval—time granularity.

Let us assume that

$t = k\Delta t$, where, $k \in \{0, 1, \ldots, k^{\max}\}$, and $k^{\max} = \left\lfloor \frac{t^{\max}}{\Delta t} \right\rfloor$.

Now we can rewrite the original model (2.36) in the following way:

$$\textit{Minimize} \quad w^{\max} \tag{2.38}$$

$$\textit{Subject to}$$

$$w^{\max} \geq \sum_{i \in I} w_i \left(q_i^{\min} + x_i^{\max}\left(1 - \frac{k\Delta t - \varphi_i}{h_i} + \left\lfloor \frac{k\Delta t - \varphi_i}{h_i} \right\rfloor\right)\right), \quad k = 0, 1, \ldots, \left\lfloor \frac{t^{\max}}{\Delta t} \right\rfloor \tag{2.39}$$

$$w^{\max} \geq 0 \tag{2.40}$$

$$0 \leq \varphi_i < h_i \quad i \in I \tag{2.41}$$

Still there is a non-linearity due to the floor function $\lfloor x \rfloor$. The usual way to model the floor function $\lfloor x \rfloor$ is to replace it with an integer variable y (StackExchange Network 2019) such that:

$$x - 1 < y \leq x. \tag{2.42}$$

To replace the strict inequality in (2.42) with a non-strict one and to provide numerical stability at the same time, we introduce a small number $\varepsilon > 0$ such that

$$x - 1 + \varepsilon \leq y \leq x, \text{or}$$

$$\varepsilon \leq 1 - x + y \leq 1$$

Replacing every term $\left\lfloor \frac{k\Delta t - \varphi_i}{h_i} \right\rfloor$ in (2.39) with an integer variable y_{ik} we rewrite the model (2.38–2.41) as:

$$Minimize \quad w^{max} \tag{2.43}$$

$$Subject\ to \quad w^{max} \geq \sum_{i \in I} w_i \left(q_i^{min} + x_i^{max} \left(1 - \frac{k\Delta t - \varphi_i}{h_i} + y_{ik} \right) \right) k = 0, 1, \ldots, \left\lfloor \frac{t^{max}}{\Delta t} \right\rfloor \tag{2.44}$$

$$\varepsilon \leq 1 - \frac{k\Delta t - \varphi_i}{h_i} + y_{ik} \leq 1, \quad \forall i, k \tag{2.45}$$

$$w^{max} \geq 0 \tag{2.46}$$

$$0 \leq \varphi_i \leq h_i \quad i \in I \tag{2.47}$$

$$y_{ik} \in \{-1, 0, 1, 2, \ldots\} \quad i \in I, \quad k = 0, 1, \ldots, \left\lfloor \frac{t^{max}}{\Delta t} \right\rfloor \tag{2.48}$$

Remark A soft inequality in (2.41) corresponds to a strong one with a minor tolerance on the right-hand side. Also, you can safely replace it on a strong one (2.47).

Let us make two substitutions:

$$z_{ik} = y_{ik} + 1 \tag{2.49}$$

$$u_{ik} = z_{ik} - \frac{k\Delta t - \varphi_i}{h_i} \tag{2.50}$$

The problem reformulation looks like:

$$Minimize \quad w^{max} \tag{2.51}$$

$$Subject\ to \quad w^{max} \geq \sum_{i \in I} w_i \left(q_i^{min} + x_i^{max} u_{ik} \right), \quad k = 0, 1, \ldots, \left\lfloor \frac{t^{max}}{\Delta t} \right\rfloor \tag{2.52}$$

$$h_i z_{ik} - h_i u_{ik} + \varphi_i = k\Delta t, \quad i \in I, \quad k = 0, 1, \ldots, \left\lfloor \frac{t^{max}}{\Delta t} \right\rfloor \tag{2.53}$$

$$w^{max} \geq 0 \tag{2.54}$$

$$0 \le \varphi_i \le h_i, \quad i \in I \tag{2.55}$$

$$\varepsilon \le u_{ik} \le 1, \quad i \in I, \quad k = 0, 1, \ldots, \left\lfloor \frac{t^{\max}}{\Delta t} \right\rfloor \tag{2.56}$$

$$z_{ik} \in \{0, 1, 2, \ldots\}, \quad i \in I, \quad k = 0, 1, \ldots, \left\lfloor \frac{t^{\max}}{\Delta t} \right\rfloor \tag{2.57}$$

Unfortunately, this new formulation (2.51–2.57) has a flaw. The phase shifts tend to be assigned with an increment εh_i to aim for the lowest values of $\{z_{ik}\}$, and as a result, the peaks of the inverse sawtooth functions are never reached. That is why along with the time discretization we must also discretize the phase shifts.

Time and Phase Shifts Discretization

Let us force the phase shifts to be multiple of the elementary time interval Δt. Designate φ_i^d as an integer variable, $\varphi_i^d \in \{0, 1, 2, \ldots, \lfloor \frac{h_i}{\Delta t} \rfloor\}$, so that $\varphi_i = \varphi_i^d \Delta t, \ i \in I$.

The model is getting the final form as the following:

$$Minimize \quad w^{\max} \tag{2.58}$$

$$Subject\ to \quad w^{\max} \ge \sum_{i \in I} w_i \left(q_i^{\min} + x_i^{\max} u_{ik} \right), \quad k = 0, 1, \ldots, \left\lfloor \frac{t^{\max}}{\Delta t} \right\rfloor \tag{2.59}$$

$$h_i z_{ik} - h_i u_{ik} + \varphi_i = k \Delta t, \quad i \in I, \quad k = 0, 1, \ldots, \left\lfloor \frac{t^{\max}}{\Delta t} \right\rfloor \tag{2.60}$$

$$\varphi_i = \varphi_i^d \Delta t, \quad i \in I \tag{2.61}$$

$$w^{\max} \ge 0 \tag{2.62}$$

$$0 \le \varphi_i \le h_i, \quad i \in I \tag{2.63}$$

$$\varepsilon \le u_{ik} \le 1, \quad i \in I, \quad k = 0, 1, \ldots, \left\lfloor \frac{t^{\max}}{\Delta t} \right\rfloor \tag{2.64}$$

$$z_{ik} \in \{0, 1, 2, \ldots\}, \quad i \in I, \quad k = 0, 1, \ldots, \left\lfloor \frac{t^{\max}}{\Delta t} \right\rfloor \tag{2.65}$$

$$\varphi_i^d \in \left\{0, 1, 2, \ldots, \left\lfloor \frac{h_i}{\Delta t} \right\rfloor \right\}, \quad i \in I \tag{2.66}$$

The above model (2.58–2.66) is an MIP model. It includes five groups of variables:

A single continuous variable $w^{\max}$,

- A vector of continuous variables $\boldsymbol{\varphi} = (\varphi_1, \varphi_2, \ldots, \varphi_m)^T$,
- A matrix of continuous variables $\{u_{ik}\}$,
- A matrix of integer variables $\{z_{ik}\}$,
- A vector of integer variables $\boldsymbol{\varphi}^d = \left(\varphi_1^d, \varphi_2^d, \ldots, \varphi_m^d\right)^T$.

The total number of integer variables is $m\left(\left\lfloor \frac{t^{\max}}{\Delta t} \right\rfloor + 2\right)$.

The model (2.58–2.66) can be computationally intensive due to potentially many integer variables $\{z_{ik}\}$ and $\boldsymbol{\varphi}^d$. A commercial solver, such as CPLEX (IBM ILOG CPLEX Optimization Studio 2017), Xpress (FICO Xpress Optimizer n.d.), and Gurobi (Gurobi Optimizer Reference Manual 2020), can tackle MIP problems of a reasonable dimension.

Model Discussion

Discretization introduces potential errors in the results. Naturally, the finer the optimization space, the lower the discretization error becomes. However, this refinement accompanies a rise in computational expenses. Consequently, obtaining an estimate of this discretization error would prove invaluable.

The second remark pertains to model-specific cuts. The objective is to align the Linear Programming (LP) relaxation as closely as possible with the convex hull of the feasibility set.

In addition, the third consideration involves employing a heuristic to offer an initial starting point for the Mixed-Integer Programming (MIP) algorithms, encompassing cuts, and the branch-and-bound method.

Let us consider a small example.

Example 2.11 WSSP. The input data are in the following table (Table 2.23):

Solutions of the continuous model with the time interval Δt as a parameter are in the table below (Table 2.24):

I present the results for $\Delta t = 0.05$ in the following graph, Fig. 2.21.

Starting at Δt of 0.100, we observe that the optimal value remains constant at 377.000, even as we further discretize time. Therefore, we can confidently conclude that the optimal value for the original problem is 377.000.

Another observation from this example is that we must pick the discretization in such way that every period h_i, $i \in I$ should be multiple of Δt, so that all peak arguments will belong to the set of discreet points $\{k\Delta t, \ k = 0, \pm 1, \pm 2, \ldots\}$.

Table 2.23 Input data for WSSP

Product	q^{min}	x^{max}	h	w	t^{max}
A	20	120	9	1	9
B	30	180	9	1	
C	25	150	9	1	

Table 2.24 Solutions of the WSSP continuous model for varying parameter Δt

Δt	φ_1	φ_2	φ_3	w^{max}	Number of integer variables	SCIP running time[a] (s)	Gurobi 10.02 running time[a] (s)
1.000	7.0	5.0	1.0	385.000	33	0.3	0.1
0.500	3.0	6.5	0.5	378.333	60	0.6	0.2
0.100	0.9	7.5	3.9	377.000	276	11.4	2.0
0.050	8.45	6.05	2.45	377.000	546	33.0	9.9

[a]The program ran on Intel(R) Core (TM) i5-10210U CPU @ 1.60GHz

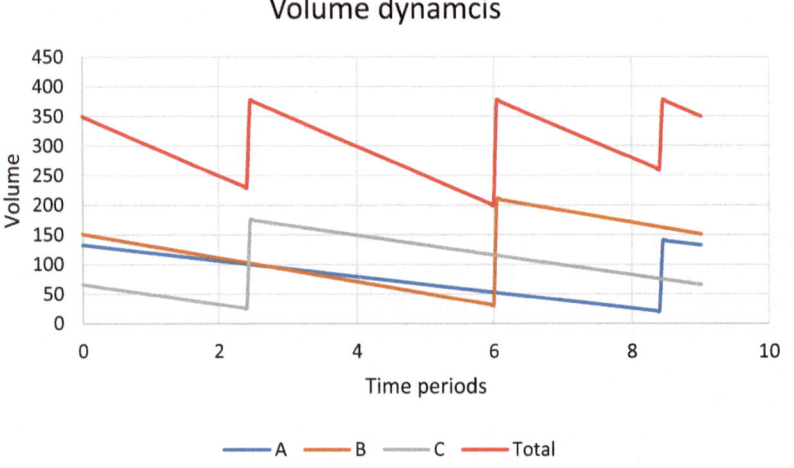

Fig. 2.21 Continuous model: volume dynamics for $\Delta t = 0.05$

Lastly, we can observe that the number of integer variables is roughly proportional to the $1/_{\Delta t}$. Undoubtedly, the number of integer variables plays a crucial role in determining the MIP performance.

We explore several ways to improve the model performance.

- Planning horizon reduction,
- Periods reduction,
- Lower bound estimate,
- Model-specific cuts,
- Model implementation.

Planning Horizon Reduction

Theorem 2.5 If all periods $\mathbf{h} = (h_1, h_2, \ldots, h_m)^T$ are rational, then the shortest planning horizon is $t^{\max} = LCM(h_1, h_2, \ldots, h_m)$, where LCM is the least common multiplier.

This theorem allows to assign the planning horizon to its minimum.

If we have two items with $h_1 = 6$ and $h_2 = 8$, then $LCM(h_1, h_2) = 24$. So, it is sufficient to run the model with $t^{\max} = 24$.

Appendix B presents a proof of the theorem.

Corollary. In the case of identical periods: $h_1 = h_2 = \ldots h_m = h_0$, the shortest planning horizon is

$$t^{\max} = h_0.$$

Periods Reduction

Theorem 2.6 Let us designate $\mathbf{P} = (\mathbf{q}^{\min}, \mathbf{x}^{\max}, \mathbf{h}, \mathbf{w})$ as the problem (2.36); $F(\mathbf{P})$ as the optimal value of problem $\mathbf{P}$; $\varphi(\mathbf{P})$ as an optimal solution to the problem $\mathbf{P}$; and $\lambda > 0$ as a factor.

Then the two problems $\mathbf{P} = (\mathbf{q}^{\min}, \mathbf{x}^{\max}, \mathbf{h}, \mathbf{w})$ and $\mathbf{P}_\lambda = (\mathbf{q}^{\min}, \mathbf{x}^{\max}, \lambda\mathbf{h}, \mathbf{w})$ have the same optimal value and proportional solutions, i.e., $F(\mathbf{P}_\lambda) = F(\mathbf{P})$ and $\varphi(\mathbf{P}_\lambda) = \lambda\varphi(\mathbf{P})$.

For example, if $\mathbf{h} = (30, 50, 70)^T$ in the problem $\mathbf{P} = (\mathbf{q}^{\min}, \mathbf{x}^{\max}, \mathbf{h}, \mathbf{w})$, then it makes sense to run the problem $\mathbf{P}_\lambda$ where $\lambda = 0.1$ and $\lambda\mathbf{h} = (3, 5, 7)^T$.

This theorem allows us to run the model with smaller values of replenishment cycles ($\lambda < 1$). However, we should be aware of the time discretization error. Appendix B presents a proof of the theorem.

Lower Bound Estimate

Conjecture. For any problem $\mathbf{P} = (\mathbf{q}^{\min}, \mathbf{x}^{\max}, \mathbf{h}, \mathbf{w})$ consider a problem $\mathbf{P}' = (\mathbf{q}^{\min}, \mathbf{x}^{\max}, h_0\mathbf{1}, \mathbf{w})$ with any replenishment cycle h_0 identical for all items, where $\mathbf{1} = (1, 1, \ldots, 1)^T$. We state that the problem $\mathbf{P}'$ provides a lower bound for the original problem $\mathbf{P}$, i.e., $F(\mathbf{P}) \geq F(\mathbf{P}')$.

The computation experiments demonstrate the validity of this conjecture. Incorporating the lower bound in the model improves the performance. In some instances, we observe performance improvement as much as 33%.

Model-Specific Cuts

The model-specific cuts aim to make an LP relaxation as close as possible to the convex hull. We have experimented with different cuts. The most promising ones are the following:

$$z_{ik} \geq z_{ik-1}, \quad i \in I, \quad k = 1, \ldots, \left\lfloor \frac{t^{\max}}{\Delta t} \right\rfloor. \tag{2.67}$$

We derive these cuts from constraints (2.45):

$$z_{ik} \geq \frac{k\Delta t - \varphi_i}{h_i} = \frac{(k-1)\Delta t - \varphi_i}{h_i} + \frac{\Delta t}{h_i} \geq z_{ik-1} + \frac{\Delta t}{h_i} \geq z_{ik-1}$$

The (2.67) cuts reduce the running time by 27% on large problems. Another model-specific cuts:

$$z_{ik} \leq z_{ik-1} + 1, \quad i \in I, \quad k = 1, \ldots, \left\lfloor \frac{t^{\max}}{\Delta t} \right\rfloor \tag{2.68}$$

We derive these cuts are from constraints (2.45) as well:

$$z_{ik} \leq \frac{k\Delta t - \varphi_i}{h_i} + 1 = \frac{(k-1)\Delta t - \varphi_i}{h_i} + \frac{\Delta t}{h_i} + 1 = z_{ik-1} + \frac{\Delta t}{h_i} + 1.$$

Considering integrality of $\{z_{ik}\}$ we come up with (2.68).

Interestingly, adding the latest cuts (2.68) does not significantly improve performance.

Model Implementation

The model is implemented in Python 3 using additional libraries from Google OR-Tools. Its website (Google OR-Tools n.d.) says: "OR-Tools is an open-source software suite for optimization, tuned for tackling the world's toughest problems in vehicle routing, flows, integer and linear programming, and constraint programming." Furthermore, later: "After modeling your problem in the programming language of your choice, you can use any of a half dozen solvers to solve it: commercial solvers such as Gurobi or CPLEX, or open-source solvers such as SCIP, GLPK, or Google's GLOP and award-winning CP-SAT."

The modeling language for declaring variables, setting up constraints, and objective functions is simple and requires only a couple of additional Python libraries:

```
from ortools.linear_solver import pywraplp
from ortools.init import pywrapinit
```

A single line of code picks a solver among a wide variety (of course a license file is required for commercial solvers). Since our model is MIP, we have selected SCIP—a robust integer programming open-source solver.

```
solver = pywraplp.Solver.CreateSolver('SCIP')
```

2.5.2 Discrete Warehouse Storage Space Model

As presented in its form (2.36), the initial warehouse storage space model follows a continuous approach, anticipating an analytical solution. In contrast, the subsequent model presented below takes on a discrete form and is slightly more comprehensive than its continuous counterpart. This discrete model explicitly incorporates factors such as product demand, maximum departures for each period, and warehouse capacity into its considerations.

Input Data

I is the set of products.
$t^{\max}$ is the planning horizon.
Δt is a time period.
T is the planning horizon that includes several periods, $T = \{1, 2, \ldots, n\}$, where

$$n = \left\lfloor \frac{t^{\max}}{\Delta t} \right\rfloor + 1.$$

$q_i^{\min}$ is the safety stock of product $i \in I$.
$x_i^{\max}$ is the maximum amount of product $i \in I$ arrived to the warehouse in period $t \in T$.
$y_i^{\max}$ is the maximum amount of product $i \in I$ departed from the warehouse in period $t \subset T$.
w_i is the volume of a unit of product $i \in I$.
h_i is the replenishment cycle—time between two arrivals of product $i \in I$.

Variables

q_{it} = amount of product $i \in I$ in a storage in period $t \in \{0\} + T$.
x_{it} = actual amount of product $i \in I$ arrived to the warehouse in period $t \in T$.

y_{it} = actual amount of product $i \in I$ departed from the warehouse in period $t \in T$.

z_{it} = indication of arrival of product $i \in I$ in period $t \in T$; $z_{it} = 1$ if the arrival happens, and $z_{it} = 0$, otherwise.

$w^{\max}$ = maximum (over time) of the total inventory volume of the stored items.

Preprocessing

Number of periods in the replenishment cycle:

$$n_i^h = \frac{h_i}{\Delta t}, \quad i \in I$$

Demand for every time period:

$$d_i = \frac{\Delta t}{h_i} x_i^{\max}, \quad i \in I$$

It means the total demand of product $i \in I$ during the replenishment cycle equals to the arrival $x_i^{\max}$.

Remark It is advisable to pick Δt as a common divisor of $\{h_i\}$.

Constraints

Inventory dynamics:

$$q_{it} = q_{i,t-1} + x_{it} - y_{it}, \quad i \in I, t \in T \tag{2.69}$$

Arrival to the warehouse:

$$x_{it} = x_i^{\max} z_{it}, \quad i \in I, t \in T \tag{2.70}$$

Departure from the warehouse:

$$y_{it} = d_i, \quad i \in I, t \in T \tag{2.71}$$

At most one arrival in any period equal to the replenishment cycle:

$$\sum_{\tau=t}^{t+n_i^h} z_{i\tau} \le 1, \quad i \in I, t = 1, n - n_i^h \tag{2.72}$$

Maximum of the total volume across all periods:

$$w^{\max} \geq \sum_{i \in I} w_i q_{it}, \quad t \in T \tag{2.73}$$

Inventory lower bound:

$$q_{it} \geq q_i^{\min} + d_i, \quad i \in I, t \in T \tag{2.74}$$

Explanation: At first glance, the lower bound of this inventory might not be readily apparent. Nevertheless, it effectively represents the scenario wherein the safety stock remains untouched, particularly during the period just before replenishment occurs. It would be incorrect to omit the demand from the right-hand side, as shown by the expression: $q_{it} \geq q_i^{\min}$, $i \in I, t \in T$.

Inventory upper bound:

$$q_{it} \leq q_i^{\min} + x_i^{\max}, \quad i \in I, t \in T \tag{2.75}$$

Non-negativity bounds:

$$x_{it} \geq 0, \quad i \in I, t \in T \tag{2.76}$$

$$y_{it} \geq 0, \quad i \in I, t \in T \tag{2.77}$$

$$q_{it} \geq 0, \quad i \in I, t \in T \tag{2.78}$$

$$w^{\max} \geq 0 \tag{2.79}$$

Binary conditions:

$$z_{it} \in \{0, 1\} \quad i \in I, \quad t \in T \tag{2.80}$$

Objective function—minimize the maximum of the total inventory volume:

$$Minimize \quad w^{\max} \tag{2.81}$$

Model Discussion

First, the model (2.69–2.81) is dynamic. The inventory dynamics (2.69) is a discrete presentation of a differential equation:

$$\frac{dq(t)}{dt} = x(t) - y(t).$$

The dynamics make the problem more attractive with various time granularity: hour, day, week, etc. The demand comes from the forecast. It brings another dimension to the model: stochasticity.

We added inventory variables corresponding to the period 0. For this model, we have not put initial inventory constraints. So, the variables $\{q_{i0}\}$ remain free.

Second, we calculated the demand $\{d_i\}$ in preprocessing, which remains the same for every period. Though the departure variables $\{y_{it}\}$ seem unnecessary, they logically separate delivery from demand.

The arrival constraints (2.70) along with inventory dynamics constraints (2.69) present a well-known inverse sawtooth function (see Fig. 2.17).

Third, the discrete warehouse storage space model (2.69–2.81) looks more straightforward than the continuous one (2.58–2.66).

Fourth, the following model-specific cuts help to improve performance:

$$\sum_{i \in I} z_{it} \le \left\lceil \frac{m}{n} \right\rceil, \quad t \in T,$$

where m is the number of products, and n is the number of periods in the planning horizon.

These cuts aim to distribute replenishments of various products across time.

Model Implementation

As the previous continuous model, the discrete model is implemented in Python 3 using additional libraries from Google OR-Tools.

Appendix C provides the Python code for the model, adopting an object-oriented design approach. The central class, `OptInvDisc`, encompasses the following functionalities:

- Constructor,
- Routine for reading a configuration file (as Table 1.1),
- Routine for reading an input file (as Table 2.23),
- Class `Solution`, which takes the solver output and converts it to the Python variables.
- The model implementing multi-criteria optimization in a lexicographic fashion,

  - Primary ranking criterion—minimizing the maximum of total inventory volume,
  - Secondary ranking criterion—maximizing the minimum of total inventory volume,

- Routine for writing solutions,
- Full program execution,
- Destructor.

Remark The inclusion of the second criterion illustrates the implementation of the lexicographic optimization and accumulation of multiple optimal solutions as objects of the class `Solution`.

For this purpose, the following constraints complement the model:
Minimum of the total volume across all periods:

$$w^{\min} \le \sum_{i \in I} w_i q_{it}, \quad t \in T,$$

where an additional variable $w^{\min}$ is the minimum (over time) of the total inventory volume of the stored items.

The second objective function—maximize the minimum of the total inventory volume:

$$Maximize \quad w^{\min}.$$

Model Testing

We tested the discrete model on the same data set as in Table 2.23. The solutions of the discrete model with the discretization time Δt as a parameter are in the table below (Table 2.25):

I present the results in the following graph, Fig. 2.22.

It is interesting to compare the results of two models: the continuous and the discrete.

Firstly, it is worth noting that both models, as we anticipate, yield an identical optimal value of 377.000.

Secondly, we can discern a superior performance from the continuous model.

Table 2.25 Solutions of the WSSP discrete model for varying parameter Δt

Δt	φ_1	φ_2	φ_3	$w^{\max}$	Number of binary variables	SCIP running time[a] (s)	Gurobi 10.02 running time[a] (s)
1.000	7.0	1.0	4.0	385.000	27	0.7	0.1
0.500	5.0	8.5	2.5	378.333	54	2.9	0.1
0.100	8.6	3.2	6.2	377.000	270	66.6	17.2
0.050	4.8	8.4	2.4	377.000	540	>1 h	288.8

[a]The program ran on Intel(R) Core (TM) i5-10210U CPU @ 1.60GHz

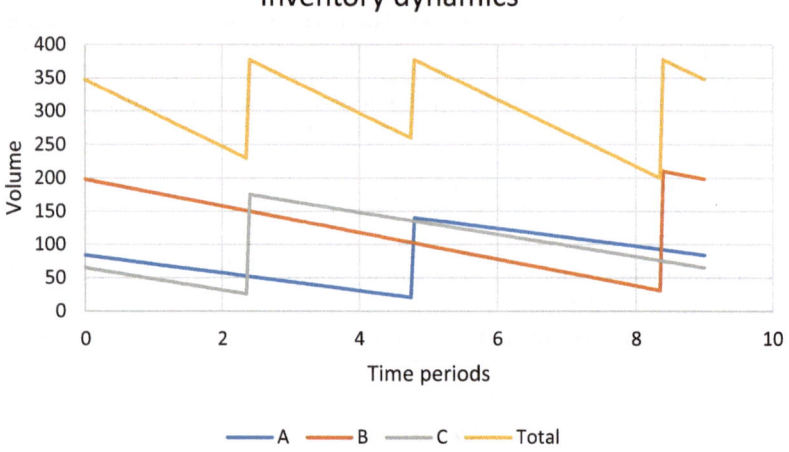

Fig. 2.22 Discrete model: volume dynamics for $\Delta t = 0.05$

2.5.3   Heuristic Algorithm

The heuristic below is a sequential greedy algorithm that finds a phase shift for every product.

Step 1. Sort products in the non-increasing order of the volume peaks $w_i\left(q_i^{\min} + x_i^{\max}\right)$.

Without loss of generality let us assume that

$$w_1\left(q_1^{\min} + x_1^{\max}\right) \geq \cdots \geq w_m\left(q_m^{\min} + x_m^{\max}\right).$$

Step 2. Initialize the first product data:

$$\varphi_1 = 0;$$

$$w(t) = w_1\left(q_1^{\min} + x_1^{\max}\left(1 - \frac{t - \varphi_1}{h_1} + \left\lfloor \frac{t - \varphi_1}{h_1} \right\rfloor\right)\right), \quad t \in [0, t^{\max}];$$

$$w^{\max} = \max {}_t\{w(t)\} = w_1\left(q_1^{\min} + x_1^{\max}\right).$$

Step 3. Loop on all remaining products, $k = 2, 3, \ldots, m$.
Step 3.1. Select $\varphi_k \in [0, h_k)$ by finding

$$w^{\max} = \min {}_{\varphi_k} \max {}_t\left\{w(t) + w_k\left(q_k^{\min} + x_k^{\max}\left(1 - \frac{t - \varphi_k}{h_k} + \left\lfloor \frac{t - \varphi_k}{h_k} \right\rfloor\right)\right)\right\}$$

Step 3.2. Update $w(t)$:

$$w(t) := w(t) + w_k \left(q_k^{\min} + x_k^{\max} \left(1 - \frac{t - \varphi_k}{h_k} + \left\lfloor \frac{t - \varphi_k}{h_k} \right\rfloor \right) \right), \quad t \in [0, t^{\max}]$$

Step 4. Output $w^{\max}$ and $\boldsymbol{\varphi} = (\varphi_1, \varphi_2, \dots, \varphi_m)^T$.
End.

This heuristic guarantees feasibility. The heuristic's complexity is $O(mn)$, where m is the number of products, and n is the number of periods caused by discretization of the time axis.

In the latest Example 2.11, the heuristic run gives the following results:

$$\varphi_A^{opt} = 5.5, \varphi_B^{opt} = 0, \varphi_C^{opt} = 4.1;$$

and

$$w^{\max} = 395.333.$$

The result is in a 5% deviation from the optimal value.

2.5.4   Conclusions

The Warehouse Storage Space Problem (WSSP) aims to minimize the required storage capacity within a warehouse. We have devised two distinct models: one based on an analytical formula (continuous) and the other on inventory dynamics (discrete). I initially developed the first model and later uncovered the second. I want to emphasize again the inherent value of employing a multi-modeling paradigm.

The execution time of both models is heavily contingent on the granularity of time discretization. Incorporating the lower bound derived from the most plausible assumption (continuous model), along with model-specific cuts for both variants, notably enhances the performance of the Mixed-Integer Programming (MIP) approach.

Furthermore, the suggested greedy heuristic exhibits remarkable speed and furnishes reasonably accurate approximate solutions.

2.6   Unit Commitment Problem

The Unit Commitment Problem (UCP) presents a complex challenge within the electricity market's supply chain domain. The electricity supply originates from various generators, including thermoelectric, nuclear, hydro, wind, and solar systems. These generators feed into transmission and distribution lines, enabling the efficient delivery of electricity to end consumers. The operational scope often encompasses a substantial geographic area, potentially spanning multiple states.

The electricity supply chain differs from traditional discrete product supply chains by the instantaneous nature of electricity consumption. Therefore, a delicate equilibrium between supply and demand must be continuously maintained. While energy storage solutions exist (large-scale electric batteries become very popular recently to provide the grid stability), they operate on a significantly smaller scale than their counterparts in traditional supply chains.

Generation reserves represent electricity supplies that are not currently active but can be rapidly mobilized in the event of power fluctuations or unforeseen generation losses. Two primary types of reserves hold significance: regulating reserves and contingency reserves.

Regulating reserves represent capacity drawn from supply sources capable of promptly increasing (ramping up or incrementing) or decreasing (ramping down or decrementing) their output within seconds, responding to control signals from the management system.

Contingency reserves represent capacity sourced from suppliers that can offset unplanned shutdowns of significant generators or transmission lines, preserving system equilibrium. While certain generators may not produce electricity actively, they must be primed to initiate generation upon request. This entails online contingency reserves—where a generating unit is dispatched but inactive (akin to a car in neutral gear)—and offline contingency reserves—where a generating unit is in standby mode, ready for activation.

Thus, a holistic approach encompassing power generation and reserves is imperative (Carlson et al. 2012).

The scheduling horizon for generator management varies. The Real-Time (RT) dispatch problem addresses a scheduling horizon spanning a few minutes. In this context, online generators remain operational without shutdown, and new generators are not dispatched (started up). When the scheduling horizon spans hours or days, the challenge becomes the Unit Commitment Problem, termed the Day Ahead (DA) problem. Given that each generator requires considerable time (hours) to initiate the operation, the DA problem assumes pivotal importance within the broader scheduling framework.

We will consider a practical UCP incorporating reserves. The verbal problem statement reads: "Given electricity demand forecasts and reserve prerequisites, schedule the operation of each potential generator to satisfy demand and reserve requirements in every interval, minimizing cumulative costs."

2.6.1   Mathematical Model

Let us consider a practical UCP presented as a part of the Edelman Award nomination by Midwest-ISO, Alstom Grid (GE now), Paragon Decision Technology (AIMMS now), and Utilicast.

Sets and Parameters

$I = \{1, 2, \ldots, |I|\}$ is the set of power-generating units i.

Remark We include all power-generating units belonging to the same grid. The actual number might be hundreds.

$L = \{1, 2, \ldots, |L|\}$ is the set of critical power transmission lines l.

Remark A critical power line operates close to its throughput. The actual number are tens.

$T = \{1, 2, \ldots, |T|\}$ is the set of periods t—the scheduling horizon.

Remark For the UCP, this set is $|T| = 24$ or 32 hours if the period length is an hour.

δ_t is the length [hours] of period $t \in T$. In classical UCP $\delta_t = \delta$ for all periods $t \in T$.

Remark Modeling time-dependent periods is an interesting generalization of the problem. It specially makes sense for long-term planning when the demand forecast is more accurate at the beginning of the scheduling horizon and less accurate toward the end.

$J_i = \{1, 2, \ldots, |J_i|\}$ are indices for a piece-wise function approximated the "generation cost–power" curve, $i \in I$.

Remark Please see Fig. 2.24.

h_t is the end of period $t \in T$. By a recursive definition:

$$h_t = h_{t-1} + \delta_t, \quad t = 1, 2, \ldots, |T|.$$

where h_0 is the starting time of the first period.

$h_i^{\min\,up}$ $\left(h_i^{\min\,down}\right)$ is the minimum duration the generating unit $i \in I$ must remain up (down) once it started up (turned off).

$h_i^{begin-up}$ $\left(h_i^{begin-down}\right)$ is the duration the generating unit $i \in I$ was up (down) at the beginning of the scheduling horizon.

h_i^{end-up} $\left(h_i^{end-down}\right)$ is the duration the generating unit $i \in I$ is supposed to be up (down) after the end of the scheduling horizon.

$T_i^{begin-up}$ are periods with predetermined commitment of the unit $i \in I$ because of the initial minimum uptime restriction:

$$T_i^{begin-up} = \left\{ t \in T \,|\, h_i^{begin-up} + h_t - h_0 \le h_i^{\min\,up} \right\}.$$

$T_i^{begin-down}$ are periods with predetermined decommitment of the unit $i \in I$ because of the initial minimum downtime restriction:

$$T_i^{begin-down} = \left\{ t \in T \,|\, h_i^{begin-down} + h_t - h_0 \le h_i^{\min\,down} \right\}.$$

T_i^{end-up} are periods with predetermined commitment of the unit $i \in I$ because of the terminal minimum uptime restriction:

$$T_i^{end-up} = \left\{ t \in T \middle| h_i^{end-up} + h_{|T|} - h_t \leq h_i^{\min up} \right\}.$$

$T_i^{end-down}$ are periods with predetermined decommitment of the unit $i \in I$ because of the terminal minimum downtime restriction:

$$T_i^{end-down} = \left\{ t \in T \middle| h_i^{end-down} + h_{|T|} - h_t \leq h_i^{\min down} \right\}.$$

Remark Because a generating unit cannot be up and down at the same time, then at least either $T_i^{begin-up}$ or $T_i^{begin-down}$ is empty. Similar, at least either T_i^{end-up} or $T_i^{end-down}$ is empty.

[a_{it}, b_{it}] is the operating range [Mw] for the generating unit $i \in I$ during a period $t \in T$ when the generating unit i provides power only.

Remarks

- It is interesting to note that the operating ranges are time-dependent.
- Normally, $a_{it} > 0$, $\forall i, t$. This condition is a prime source of integrality in the unit commitment problem because the feasible status of a generating unit is either 0 or [a_{it}, b_{it}] (see Fig. 2.23).

[a_{it}^{rr}, b_{it}^{rr}] is the operating range [Mw] for the generating unit $i \in I$ during a period $t \in T$ when the generating unit i provides a regulating reserve.

$$0 < a_{it} \leq a_{it}^{rr} \leq b_{it}^{rr} \leq b_{it}.$$

In the sets notation: $\left[a_{it}^{rr}, b_{it}^{rr} \right] \subset [a_{it}, b_{it}]$.

μ_{ij}, η_{ij} are cost coefficients of the piece-wise approximation for the "Generation cost–Power" curve $j \in J_i$, $i \in I$.

μ_{ij} is measured in [\$/Mw] and η_{ij} is in [\$]. Fortunately, the "Generation cost–Power" curve is a convex function that can be easily modeled in the minimization problem (Fig. 2.24).

$c_{itp}^{startup}$ is the startup cost [\$] for the generating unit $i \in I$ when the generating unit starts at a period $t \in T$, while being down from a period p to a period $t - 1$. The

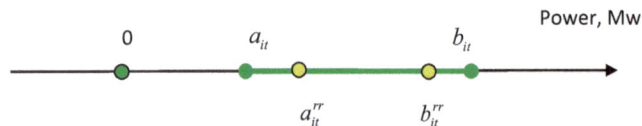

Fig. 2.23 A feasible set (in green) for a generating unit power

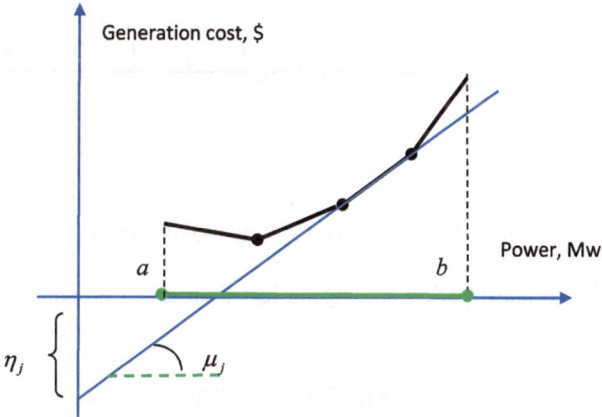

Fig. 2.24 "Generation cost–Power" function—piecewise approximation

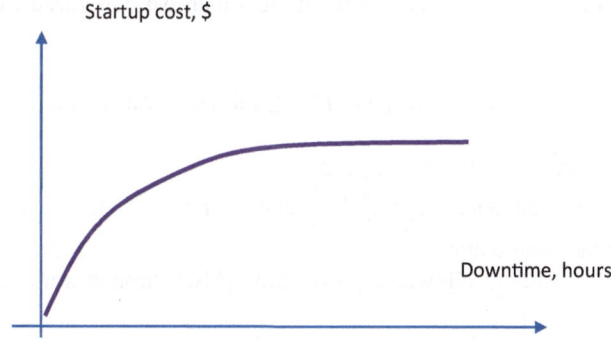

Fig. 2.25 "Startup cost–Downtime" curve

"Startup cost–Downtime" curve is in Fig. 2.25. The startup cost increases with downtime increase. We assume that:

$$c_{it1}^{startup} \geq c_{it2}^{startup} \geq \cdots \geq c_{it,t-1}^{startup}.$$

c_{it}^{rr} is the cost of regulating reserve [$/Mwh] for the generating unit $i \in I$ during a period $t \in T$.

c_{it}^{on} is the cost of online contingency reserve [$/Mwh] for the generating unit $i \in I$ during a period $t \in T$.

c_{it}^{off} is the cost of offline contingency reserve [$/Mwh] for the generating unit $i \in I$ during a period $t \in T$.

d_t is the power demand [Mw] during a period $t \in T$, $d_t \geq 0$.

d_t^{rr} is the minimum requirement for regulating reserve [Mw] during a period $t \in T$, $d_t^{rr} \geq 0$.

Fig. 2.26 Function
$e_-(t, \Delta h)$ returns p

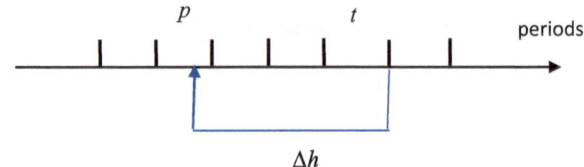

d_t^{cr} is the minimum requirement for contingency reserve [Mw] during a period
$t \in T$, $d_t^{cr} \geq 0$.

$e_-(t, \Delta h)$ is the latest period staying no less than $\Delta h > 0$ behind from a period
$t \in T$:

$$e_-(t, \Delta h) = \max\{p \in T | h_p \leq \max(h_t - \Delta h, h_0)\}.$$

Figure 2.26 illustrates function $e_-(t, \Delta h)$.

We calculate in advance $e_-\left(t, h_i^{\min up}\right)$ and $e_-\left(t, h_i^{\min down}\right)$ for $t \in T$, $i \in I$.

$e_+(t, \Delta h)$ is the earliest period staying no less than $\Delta h > 0$ ahead from a period
$t \in T$:

$$e_+(t, \Delta h) = \min\{p \in T | h_p \geq \min\left(h_t + \Delta h, h_{|T|}\right)\}$$

Figure 2.27 illustrates function $e_+(t, \Delta h)$.

We calculate in advance $e_+\left(t, h_i^{\min up}\right)$ and $e_+\left(t, h_i^{\min down}\right)$, $t \in T$, $i \in I$, for the
alternative model formulation.

$f_l^{\max}$ is the maximum allowable power flow [Mw] through a transmission line
$l \in L$, $f_l^{\max} \geq 0$.

g_{il} is the power flow along a transmission line $l \in L$ by the generating unit $i \in I$
injecting one Mw power. Since the power flow direction is important, g_{il} can be
positive or negative.

$k_i^{rr}\left(k_i^{cr}\right)$ is the portion of regulating (contingency) reserve for the generating unit
$i \in I$ considering the ramping constraints.

r_{it} is the maximum allowable ramping up or down [Mw/h] for the generating unit
$i \in I$ during a period $t \in T$.

$z_{it}^{off\,\max}$ is the maximum allowable offline contingency reserve [Mw] provided by
the generating unit $i \in I$ during a period $t \in T$.

The initial and terminal values for all variables below ($t = 0$ and $t = |T|$) are in
the list of parameters.

Variables

x_{it} = power [Mw] generated by the unit $i \in I$ during a period $t \in T$.

Fig. 2.27 Function
$e_+(t, \Delta h)$ returns p

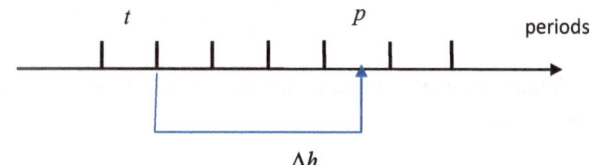

y_{it} = amount of regulating reserve [Mw] provided by the unit $i \in I$ during a period $t \in T$.

$z_{it}^{on} \left(z_{it}^{off} \right)$ = amount of contingency reserve [Mw] provided by the online (offline) unit $i \in I$ during period $t \in T$.

u_{it} = 1 if the unit $i \in I$ commits during a period $t \in T$, and 0 otherwise.

v_{it} = 1 if the unit $i \in I$ commits regulating reserve during a period $t \in T$, and 0 otherwise.

s_{it}^{up} = 1 if the unit $i \in I$ starts up during a period $t \in T$, and 0 otherwise.

s_{it}^{down} = 1 if the unit $i \in I$ shuts down during a period $t \in T$, and 0 otherwise.

σ_{it}^{gen} = generation and no-load cost [$] of the unit $i \in I$ during a period $t \in T$.

Remark *No-load cost* is associated with an online generating unit that does not supply power to the grid.

$\sigma_{it}^{startup}$ = startup cost [$] of the unit $i \in I$ during a period $t \in T$.

Remark Turning the unit on incurs the *startup cost*.

Objective Function

Minimize total cost: generation, no-load, startup, regulating and contingency reserve costs:

$$\sum_{i \in I} \sum_{t \in T} \left[\sigma_{it}^{gen} + \sigma_{it}^{startup} + c_{it}^{rr} \delta_t y_{it} + c_{it}^{on} \delta_t z_{it}^{on} + c_{it}^{off} \delta_t z_{it}^{off} \right].$$

Constraints

Power balance: Power produced must be instantaneously consumed:

$$\sum_{i \in I} x_{it} = d_t, \quad t \in T.$$

Regulating reserve requirements: Total regulating reserves must honor the minimum requirements:

$$\sum_{i\in I} y_{it} \geq d_t^{rr}, \quad t \in T.$$

Total reserve requirements: Total reserve must honor the minimum requirements:

$$\sum_{i\in I}\left(y_{it} + z_{it}^{on} + z_{it}^{off}\right) \geq d_t^{rr} + d_t^{cr}, \quad t \in T.$$

Power flow restrictions for transmission lines: The absolute value of the power flow at every critical transmission line is restricted:

$$-f_l^{\max} \leq \sum_{i\in I} g_{il} x_{it} \leq f_l^{\max} \quad l \in L, t \in T.$$

Relationship between commitment and regulating reserve commitment: Commitment is necessary for regulating reserve commitment:

$$v_{it} \leq u_{it} \quad i \in I, t \in T.$$

Dispatch, regulating and online contingency reserves relationship:

$$a_{it} u_{it} + \left(a_{it}^{rr} - a_{it}\right) v_{it} + y_{it} \leq x_{it} \leq b_{it} u_{it} + \left(b_{it}^{rr} - b_{it}\right) v_{it} - y_{it} - z_{it}^{on} \quad i \in I, t \in T.$$

Explanation. If a generating unit does not provide a regulating reserve, then $v_{it} = 0$, and

$$a_{it} u_{it} \leq x_{it} \leq b_{it} u_{it} - z_{it}^{on} \quad i \in I, t \in T$$

If a generating unit does provide a regulating reserve, then $v_{it} = 1$; $u_{it} = 1$, and

$$a_{it}^{rr} + y_{it} \leq x_{it} \leq b_{it}^{rr} - y_{it} - z_{it}^{on} \quad i \in I, t \in T.$$

Online contingency reserve works only when a unit is up ($u_{it} = 1$), and this reserve allows to dispatch the unit higher.

Regulating reserve restrictions: Half of the operating range restricts the regulating reserve:

$$y_{it} \leq 0.5\left(b_{it}^{rr} - a_{it}^{rr}\right) v_{it}, \quad i \in I, t \in T.$$

Remarks

- These regulating reserve restrictions follow immediately from the previous constraints in the form:

$$a_{it}^{rr} + y_{it} \leq b_{it}^{rr} - y_{it}, \quad i \in I, t \in T.$$

- Though these constraints do not squeeze the feasibility set, they establish a relationship between two important groups of variables: $\{y_{it}\}$ and $\{v_{it}\}$.

Offline contingency reserve: Offline contingency reserve works only when the unit is down ($u_{it} = 0$) and has the potential to be committed and generate contingency reserve power:

$$z_{it}^{off} \leq (1 - u_{it}) z_{it}^{off\ max}, \quad i \in I, t \in T.$$

Startup, shutdown, and commitment relationship:

$$s_{it}^{up} - s_{it}^{down} = u_{it} - u_{i,t-1} \quad i \in I, t \in T.$$

Explanation. If a unit starts up in a period $t \in T$, it means that $u_{it} = 1$; $u_{i,\,t-1} = 0$; and $s_{it}^{down} = 0$. If a unit shuts down in a period $t \in T$, it means $u_{it} = 0$; $u_{i,\,t-1} = 1$; and $s_{it}^{up} = 0$.

Startup and shutdown relationship: No startup and shutdown are allowed at the same time:

$$s_{it}^{up} + s_{it}^{down} \leq 1 \quad i \in I, t \in T.$$

Ramping up restrictions: Ramping up includes power change and portions of regulating and contingency reserves:

$$x_{it} - x_{i,t-1} + k_i^{rr} y_{it} + k_i^{cr} z_{it}^{on} \leq (1 - s_{it}^{up}) r_{it} \delta_t + s_{it}^{up} a_{it}, \quad i \in I, t \in T.$$

Explanation. Ramping up happens in two cases:

1. A generating unit is up and moves to a higher generating level: $s_{it}^{up} = 0$, and

$$x_{it} - x_{i,t-1} + k_i^{rr} y_{it} + k_i^{cr} z_{it}^{on} \leq r_{it} \delta_t, \quad i \in I, t \in T.$$

2. A generating unit is down and starts: $s_{it}^{up} = 1$, $x_{i,\,t-1} = 0$ reaching the lowest dispatch level:

$$x_{it} + k_i^{rr} y_{it} + k_i^{cr} z_{it}^{on} \leq a_{it}, \quad i \in I, t \in T.$$

Ramping down restrictions: Ramping down includes power change and a portion of regulating reserve:

$$x_{i,t-1} - x_{it} + k_i^{rr} y_{it} \le \left(1 - s_{it}^{down}\right) r_{it} \delta_t + s_{it}^{down} b_{it}, \quad i \in I, t \in T.$$

Explanation. Ramping down happens in two cases:

1. A generating unit is up and moves to a lower generating level: $s_{it}^{down} = 0$, and

$$x_{i,t-1} - x_{it} + k_i^{rr} y_{it} \le r_{it} \delta_t, \quad i \in I, t \in T.$$

2. A generating unit is up and shuts down: $s_{it}^{down} = 1$, $x_{it} = 0$, $y_{it} = 0$ even from the highest level of the dispatch range:

$$x_{i,t-1} \le b_{it}, \quad i \in I, t \in T.$$

Minimum uptime constraints: Once a generator is turned on it should remain operational for at least a certain amount of time, called a *minimum uptime* (Hedman et al. 2009).

$$\sum_{p=e_-\left(t,h_i^{\min up}\right)}^{t} s_{ip}^{up} \le u_{it}, \quad i \in I, t \in T.$$

Explanation: If a unit is down in time t, no startups are allowed in the previous periods inside of the minimum uptime (Fig. 2.28).

Remarks

- Due to these requirements, we cannot break up the overall unit commitment problem into several sub-problems, each for every period. Thus, all periods are tied up.
- There exists an alternative formulation of the minimum uptime constraints:

$$u_{it} - u_{i,t-1} \le u_{i\tau}, \quad \tau = t+1, \ldots, e_+\left(t, h_i^{\min up}\right), \quad t \in T.$$

However, according to the theorem proven by Rajan and Takriti (2005), the original formulation is stronger (in LP metric) than the alternative one. Our numerical experiments confirmed this theoretical result. The program with the original formulation runs about 15% faster than the alternative one.

Minimum downtime constraints: Once a generator is turned off, it should stay in this state for at least a certain amount of time, called a *minimum downtime* (Hedman et al. 2009).

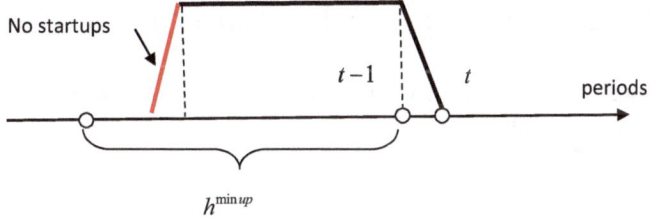

Fig. 2.28 Minimum uptime constraint illustration

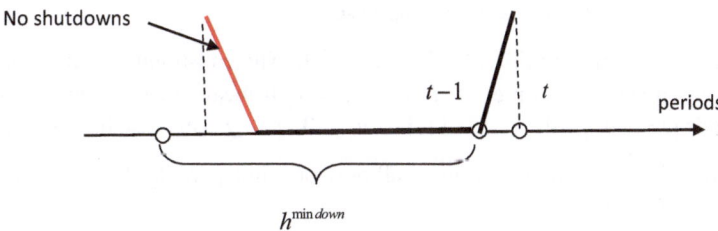

Fig. 2.29 Minimum downtime constraints illustration

$$\sum_{p=e_-\left(t,h_i^{\min down}\right)}^{t} s_{ip}^{down} \leq 1 - u_{it}, \quad i \in I, t \in T.$$

Explanation. If a unit is up in time t, no shutdowns are allowed in the previous periods inside of the minimum downtime (Fig. 2.29).

Remark There exists an alternative formulation of the minimum downtime constraints:

$$u_{i,t-1} - u_{it} \leq 1 - u_{i\tau}, \quad \tau = t+1, \ldots, e_+\left(t, h_i^{\min down}\right), \quad t \in T.$$

However, like in the case of minimum uptime constraint, the original formulation is stronger than the alternative one.

Generation and no-load cost restrictions: approximation of the convex "Generation cost–Power" curve:

$$\sigma_{it}^{gen} \geq \mu_{ij}xit + \eta_{ij}u_{it}, \quad j \in J_i, i \in I, t \in T.$$

Explanation. The "Generation cost–Power" curve is a convex function approximated by a piece-wise function presented in Fig. 2.24. A set of inequalities models this piece-wise convex function.

Startup cost restrictions: Approximation of the "Startup cost–Downtime" curve and the downtime restrictions:

$$\sigma_{it}^{startup} \geq c_{itp}^{startup} \left(s_{it}^{up} - \sum_{k=p}^{t-1} u_{ik} \right), \quad i \in I, t \in T, p = 1, \ldots, t-1.$$

Explanation. If the startup cost were not downtime dependent, then the constraints would be as simple as:

$$\sigma_{it}^{startup} = c_{it}^{startup} s_{it}^{up}, \quad i \in I, t \in T,$$

where $c_{it}^{startup}$ is a startup cost of unit $i \in I$ in time $t \in T$.

Our situation is more complex since the startup cost is a function of downtime (see Fig. 2.25). There are the following cases:

1) There is no startup in time $t \in T$, i.e., $s_{it}^{up} = 0$. The constraints are not binding.
2) There is a startup in time $t \in T$, i.e., $s_{it}^{up} = 1$. It means that the unit is down in periods $Q = \{q, q+1, \ldots, t-1\}$, i.e., $u_{it} = 0$, $t \in Q$. This case has two subcases:

    2a) $q = 1$, i.e., the unit is down in all previous time periods. The constraints boil down to the form:

$$\sigma_{it}^{startup} \geq c_{itp}^{startup}, \quad i \in I, t \in T, p = 1, \ldots, t-1.$$

    2b) $q > 1$ and $u_{i, q-1} = 1$. It means the unit shuts down in time period q. The constraints become:

$$\sigma_{it}^{startup} \geq c_{itp}^{startup}, \quad i \in I, t \in T, p \in Q, \text{and non-existed for} p \in \{1, \ldots, q-1\}.$$

Initial conditions: A unit must remain committed (uncommitted) if it was up (down) at the beginning of the scheduling horizon:

$$u_{it} = 1, \quad i \in I, t \in T_i^{begin-up}$$

$$u_{it} = 0, \quad i \in I, t \in T_i^{begin-down}.$$

Terminal conditions: A unit must remain committed (uncommitted) if it is supposed to be up (down) after the end of the scheduling horizon:

$$u_{it} = 1, \quad i \in I, t \in T_i^{end-up}$$

$$u_{it} = 0, \quad i \in I, t \in T_i^{end-down}$$

The non-negativity bounds for energy dispatch and reserves:

$$x_{it} \geq 0, \quad i \in I, t \in T$$

$$y_{it} \geq 0, \quad i \in I, t \in T$$

$$z_{it}^{on} \geq 0, \quad i \in I, t \in T$$

$$z_{it}^{off} \geq 0, \quad i \in I, t \in T$$

The startup and shutdown bounds:

$$0 \leq s_{it}^{up} \leq 1, \quad i \in I, t \in T$$

$$0 \leq s_{it}^{down} \leq 1, \quad i \in I, t \in T$$

Explanation. We declare variables s_{it}^{up} and s_{it}^{down} as continuous to reduce the number of integer (binary) variables. However, due to constraints

$$s_{it}^{up} - s_{it}^{down} = u_{it} - u_{i,t-1}, \quad i \in I, t \in T$$

these variables will take $\{0,1\}$ values in the optimal solutions.

Integrality constraints for commitment and regulating reserve commitment:

$$u_{it} \in \{0, 1\}, \quad i \in I, t \in T$$

$$v_{it} \in \{0, 1\}, \quad i \in I, t \in T.$$

Additional constraints: Please see "Model specific cuts."

The following table (Table 2.26) shows the differences between classical and real-world unit commitment problems.

2.6.2 Model-Specific Cuts

We will consider two model-specific cuts: demand-related and ramping-related cover cuts.

Table 2.26 Functional requirements for classical and real-world UC problem

	Classical UCP	Real-world UCP
Scope	Power generation.	Power generation and reserves.
"Generation cost–Power"	A linear function.	A non-linear convex function.
"Startup cost–Downtime"	Constant.	A non-linear concave function.
Periods	Constant length.	Variable length.

Demand-Related Cover Cuts

We explore "cover cuts" for the unit commitment problem. Implementing cover cuts aims to enhance the tightness of the linear programming (LP) relaxation, ultimately improving the efficiency of the original Mixed-Integer Programming (MIP) model. It is worth noting that these cuts do not eliminate any feasible solutions from the original MIP model but serve to prune a portion of the LP relaxation. M. Zhao and J. Kalagnanam initially proposed these cut techniques for addressing the unit commitment problem (Zhao and Kalagnanam 2009).

Let us remind the power balance for every period:

$$\sum_{i \in I} x_{it} = d_t, \quad t \in T.$$

Dispatch range constraints without reserves are the following:

$$a_{it} u_{it} \le x_{it} \le b_{it} u_{it}, \quad i \in I, t \in T.$$

Let us introduce a subset of generating units $I_t^0 \subset I$ such that the total capacity of the units belonging to this subset is insufficient to satisfy the energy demand. In a formal way:

$$\sum_{i \in I_t^0} b_{it} < d_t, \quad t \in T.$$

Now, the apparent logical implication is the following. Since the subset I_t^0 is insufficient to satisfy the demand, the remaining complementary subset of the units must contain at least one committed unit. Formally:

$$\exists \ i \in I \backslash I_t^0 \text{ such that } x_{it} \ge a_{it}.$$

So, the cover cuts are:

$$\sum_{i \in I \backslash I_t^0} x_{it} / a_{it} \ge 1, \quad t \in T.$$

Example 2.12 UCP. This is a trivial example with two generating units. The first unit does not have enough capacity to satisfy the demand, i.e., $b_1 < d$; therefore, the cover cut is

$$x_2 / a_2 \ge 1 \text{ or } x_2 \ge a_2.$$

Figure 2.30 illustrates this case.
It immediately implies: $u_2 = 1$; $x_2 \ge a_2$.

Fig. 2.30 Example of a
cover cut: the second unit
must be dispatched to satisfy
the demand

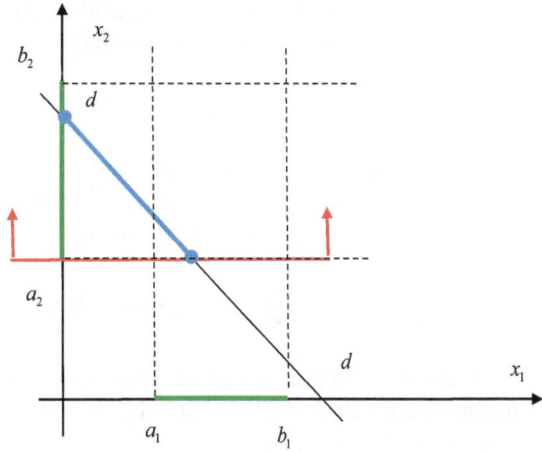

By the way, to find a promising subset I_t^0 it is sufficient to solve a knapsack problem for any period $t \in T$:

$$Maximize \quad \sum_{i \in I} z_{it}$$

$$Subject \ to \quad \sum_{i \in I} b_{it} z_{it} \le d_t(1 - \varepsilon)$$

$$z_{it} \in \{0, 1\}.$$

Here, ε is a small positive number, $\varepsilon > 0$.
Thus, we generate a solution for every period $t \in T$.

$$I_t^0 = \{i \in I | z_{it}^{opt} = 1\}.$$

Our numerical experiments with Midwest-ISO show that although LP relaxation is getting tighter, the performance has no significant gain. The explanation for this fact may be twofold. First, the generation capacity is not a restricting factor at Midwest-ISO, and second, the operating range is wide enough, so the restriction is weak.

Ramping-Related Cover Cuts

Ramping involves modifying the power generation level. The power generators we work with cannot instantaneously adjust their generation levels; it requires time. As a result, ramping can pose constraints, particularly when there is rapid growth in demand. In such instances, ramping capacity refers to the capability of online generating units to respond effectively to these quickly changing demand patterns.

If we ignore regulating reserve and consider energy only, the ramping constraints for a committed generating unit look like this:

$$- r_{it}\delta_t \leq x_{it} - x_{i,t-1} \leq r_{it}\delta_t, \quad i \in I, t \in T.$$

Now, let us consider two adjacent periods when the demand changes, say, grows fast. Let us consider a subset of generating units $I_t^1 \subset I$ that do not have sufficient ramping capacity to pick up this fast-growing demand:

$$\sum_{i \in I_t^1} r_{it}\delta_t < d_t - d_{t-1}, \quad t \in T.$$

If a feasible solution exists, at least one committed unit must be from the complementary subset $I \backslash I_t^1$. A cut like the one considered above expresses the commitment of such a unit:

$$\sum_{i \in I \backslash I_t^1} x_{it}/a_{it} \geq 1, \quad t \in T.$$

A solution to the following knapsack problem generates a subset I_t^1 for every period $t \in T$.

$$Maximize \quad \sum_{i \in I} z_{it}$$

$$Subject \ to \quad \sum_{i \in I} r_{it}\delta_i z_{it} \leq (d_t - d_{t-1})(1 - \varepsilon)$$

$$z_{it} \in \{0, 1\}.$$

Here, ε is a small positive number, $\varepsilon > 0$.

$$I_t^1 = \{i \in I | z_{it}^{opt} = 1\}.$$

This cut appears to be much stronger, at least for the data sets run at Midwest-ISO.

2.6.3   Avoiding Infeasibility

Infeasibility is an undesirable outcome. So, preventing infeasibility contributes to smooth operational flow and better overall performance.

A practical approach involves selectively relaxing certain constraints and incorporating the highest-priority infeasibility criterion into the list of objective functions (as seen in Lexicographic optimization). However, it is imperative to acknowledge that conducting an additional run can demand substantial time resources in the context of Mixed-Integer Programming (MIP).

One workaround is constructing a supplementary Linear Programming (LP) model to identify potential relaxations before executing the MIP model. The supplementary problem aims to eliminate infeasibility with minimal cost implications—seeking minor adjustments to the input data that yield feasible solutions.

To illustrate this strategy, consider a scenario where the dynamic range of unit operation conflicts with the maximum ramping rate. We pose the problem: "What is the least possible relative relaxation of the maximum ramping-up/ramping-down rates that ensure feasible dispatch across all periods?" We propose the following model.

Mathematical Model

Let us consider a single power generating unit.

Input Data

T is the set of periods, $T = \{0, 1, 2, \ldots, |T|\}$.
Without loss of generality, let us assume $\delta_t = 1$, $t \in T$.
$[a_t, b_t]$ is the dispatch range in a period $t \in T$.

Remark If the initial dispatch is given, then $a_0 = b_0$.
r^{up} is the maximum ramping up, $r^{up} > 0$.
r^{down} is maximum ramping down, $r^{down} > 0$.

Variables

x_t = generating unit dispatch in a period $t \in T$.
α = relative increase of ramp-up and ramp-down capacity.

Constraints

$$a_t \leq x_t \leq b_t, \quad t \in T$$

$$-r^{down}(1 + \alpha) \leq x_t - x_{t-1} \leq r^{up}(1 + \alpha), \quad t \in T \setminus \{0\}$$

Objective Function

Minimize α

It is interesting to note that this model is always feasible (assuming that the input data are valid: $0 < a_t \le b_t$, $t \in T$, and $r^{up} > 0$, $r^{down} > 0$). The proof is trivial: there exists a solution:

$$ x_t = a_t, \ t \in T, \text{and } \alpha = \max \left\{ \frac{\max_{t \in T \setminus \{0\}} \{|x_t - x_{t-1}|\}}{\min\{r^{up}, r^{down}\}} - 1, 0 \right\} $$

Let us consider the following example.

Example 2.13 The input data is in the table below (Table 2.27):

We conclude that the original problem, as formulated, is infeasible because the dispatch at $t = 1$ must be 120, and the subsequent dispatch of 180 at $t = 2$ cannot be reached because of a restricted rump-up of 50.

What is the minimal relaxation of the ramp-up (or ramp-down) to make the problem feasible?

Running the above model answers: $\alpha = 0.1$. So, if $r^{up} = 55$, the problem becomes feasible. Indeed, a possible dispatch solution is the following (Table 2.28):

Figure 2.31 depicts the solution.

2.6.4   Conclusions

The practical formulation of the unit commitment problem makes UCP even more complicated than its classical counterpart, encompassing the integration of binary variables for regulating reserve commitment alongside the conventional binary commitment variables.

We presented certain constraints in two optional formulations within this intricate framework. Our preference naturally leans toward formulations that induce a more stringent Linear Programming (LP) relaxation, amplifying the model's precision.

We deploy tailored model-specific cuts to heighten the LP relaxation's effectiveness and enhance performance. These rationally designed cuts compress the UCP's LP relaxation, yielding improvements in specific instances.

Table 2.27 Input data for the supplementary problem	t	a_t	b_t	r^{up}	r^{down}
	0	70	70	50	20
	1	120	150		
	2	180	200		
	3	140	160		

Table 2.28 Solution of the supplementary problem

t	x_t
0	70
1	125
2	180
3	160

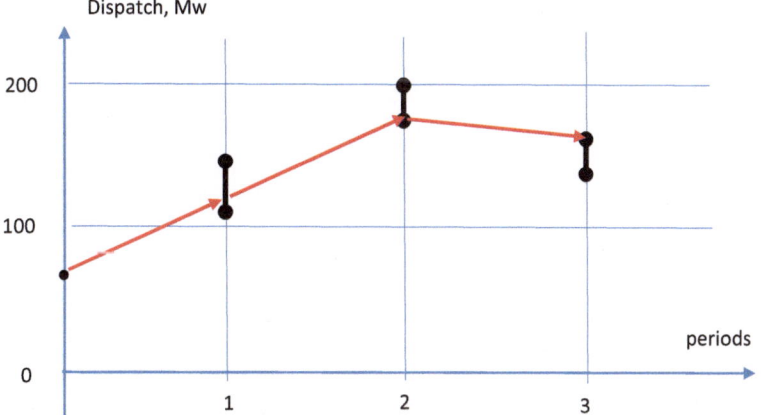

Fig. 2.31 Graphic illustration of Table 2.28

A supplementary LP model assesses the problem's potential infeasibility. This supplementary model systematically examines the feasibility of the UCP and identifies possible adjustments of the input data to avoid infeasibility.

The comprehensive project delivered substantial benefits to our client, Midwest-ISO, while earning recognition from my professional community, INFORMS, by winning a prestigious Edelman Award (Carlson et al. 2012; Zak 2011).

Correction to: Real-World Problems

Correction to:
Chapter 2 in: E. J. Zak, *How to Solve Real-world Optimization Problems*, SpringerBriefs in Operations Research, https://doi.org/10.1007/978-3-031-49838-1_2

Inadvertently, the second equation under Section "Preprocessing" on Page 72 was published with an error. This has now been rectified to read as

$$d_i = \frac{\Delta t}{h_i} x_i^{\max}, \quad i \in I$$

The equation (2.74) on Page 73 was published with an error. This has now been corrected to read as

$$q_{it} \geq q_i^{\min} + d_i, \quad i \in I, t \in T \tag{2.74}$$

The updated version of this chapter can be found at
https://doi.org/10.1007/978-3-031-49838-1_2

Concluding Remarks

A well-known adage asserts: "The gap between theory and practice is larger in practice than in theory." The interests of academic scientists and practitioners are not inherently adversarial; rather, they operate in orthogonal directions. Practitioners encounter real-world predicaments that academic scientists investigate, while advancements in theory invariably positively influence practical applications. Nevertheless, disparities persist between theoretical constructs and practical implementation, and this book highlights these disparities from a practitioner's perspective.

In Chap. 1, I have distilled practical insights from my setbacks and triumphs in addressing tangible Operational Research challenges. I further illuminate these insights in Chap. 2.

An essential theme is a profound understanding of the subject domain. I conducted on-site visits to clients in numerous instances, delving into diverse environments such as paper mills, power grid transmission and distribution control rooms, and warehouses. These immersive visits provided invaluable insights that deepened my understanding of the problems and their subsequent solutions.

Direct feedback from clients plays a pivotal role in formulating robust optimization models for real-world problems and aids in creating complementary heuristic algorithms.

Undoubtedly, infeasibility is an unwelcome outcome for an end user dealing with an optimization model. In general, the problem constraints can be made "soft." Alternatively, auxiliary models can address data compatibility issues directly. A relevant illustration is the relative ramping relaxation model, developed to address the unit commitment problem.

In rare cases, a practical problem may have a single objective function, but in most cases, you will engage in multi-criteria optimization. For practical implementation, comprehension, and control, I employ lexicographic optimization as an accessible method for handling multiple criteria.

Since every model mirrors the real world, the end-user will appreciate having several candidates for the final solutions. Mathematically, optimal points within the

E. J. Zak, *How to Solve Real-world Optimization Problems*, SpringerBriefs in
Operations Research, https://doi.org/10.1007/978-3-031-49838-1

optimal face can be explored using specialized algorithms or lexicographic optimization. Heuristics provide near-optimal solutions that remain invaluable to end users.

Defining the boundaries of a problem domain and its interconnected or supplementary components is paramount in system analysis. A notable example in my modeling repertoire involves the discovery of the Skiving Stock Problem as a complement to the Cutting Stock Problem. This discovery profoundly influenced the architecture of the encompassing software suite addressing both challenges.

A diverse array of optimization models emerges for the same optimization problem. Crafting an intelligent model demands time and heightened expertise, yet the investment pays off in more effective deployment, maintenance, and support. For instance, although the discrete model for the Warehouse Storage Space Problem is straightforward and suitable for practical use, the continuous model surpasses it in terms of performance. However, it is important to emphasize that this does not diminish the value of the discrete model; rather, it can serve as a valuable complementary approach.

It is essential to acknowledge that the array of practical insights outlined in Chap. 1 is not exhaustive. Topics such as parallel computing, online optimization, and AI-based models remain unexplored. This book is not a comprehensive reference manual but rather a practitioner's compass navigating crucial terrain in modeling real-world optimization problems.

Appendices

Appendix A: Bin-Packing Model Solution

The table below (Table A.1) shows a bin-packing model solution of Example 2.1.

Appendix B: The Theorem Proofs

Proof of Theorem 2.1

First, notice that the objective value of the LP relaxation of the bin-packing model is $\mathbf{w}^T\mathbf{b}/w^{stock}$.

A solution for the LP relaxation of the bin-packing problem is the following. We create m bin types with complete filling:

$$a_{ii} = w^{stock}/w_i, \quad i \in I \text{ and } a_{ij} = 0, \quad i \in I, j \in I, i \neq j.$$

Each bin type i is repeated b_i/a_{ii} times, $i \in I$, totaling

$$\sum_{i\in I} b_i/a_{ii} = \left(\sum_{i\in I} w_i b_i\right)/w^{stock} = \mathbf{w}^T\mathbf{b}/w^{stock} \text{ bins.}$$

Since this solution assumes complete filling, all other solutions can have complete filling (no trim loss) or partial filling. That is why $\mathbf{w}^T\mathbf{b}/w^{stock}$ is the least value among all CSP solutions. Consequently, the integrality gap of the bin-packing model is the largest. Q.E.D.

© The Author(s), under exclusive license to Springer Nature Switzerland AG 2024
E. J. Zak, *How to Solve Real-world Optimization Problems*, SpringerBriefs in
Operations Research, https://doi.org/10.1007/978-3-031-49838-1

Table A.1 Bin-packing model solution of Example 2.1

Order	Width	1	2	3	4	5	6	7	8	9	10	11	12
A	16.5	5	0	0	0	0	2	0	0	0	8	2	0
B	17	0	0	0	1	1	0	0	0	2	4	3	0
C	18	0	0	3	2	0	0	0	3	0	0	5	0
D	23	1	0	0	0	1	7	0	6	7	0	0	0
E	25.75	0	1	0	0	5	0	1	0	0	0	1	1
F	29	1	6	5	5	0	0	6	0	0	0	0	6
G	31.25	2	0	0	0	1	0	0	0	0	0	0	0
H	32	0	0	0	0	0	0	0	0	0	0	0	0
	Space	197	199.75	199	198	200	194	199.75	192	195	200	199.75	199.75

Order	Width	13	14	15	16	17	18	19	20	21	22	23	24
A	16.5	0	0	0	0	1	0	0	0	0	0	0	0
B	17	0	1	7	0	0	5	3	8	1	0	0	0
C	18	2	1	0	4	0	0	0	0	0	2	1	0
D	23	7	3	0	0	0	5	2	0	1	6	0	2
E	25.75	0	0	2	0	6	0	4	0	0	1	1	0
F	29	0	0	1	0	1	0	0	0	0	0	0	2
G	31.25	0	0	0	0	0	0	0	2	0	0	5	0
H	32	0	3	0	4	0	0	0	0	5	0	0	3
	Space	197	200	199.5	200	200	200	200	198.5	200	199.8	200	200

Order	Width	25	26	27	28	29	30	31	32	b	Placed
A	16.5	0	1	0	1	11	0	0	1	32	32
B	17	0	1	0	0	0	0	0	10	47	47
C	18	1	0	1	0	1	0	0	0	26	26
D	23	0	5	0	0	0	3	8	0	64	64
E	25.75	0	2	1	6	0	0	0	0	32	32
F	29	3	0	0	1	0	0	0	0	37	37
G	31.25	3	0	5	0	0	4	0	0	22	22
H	32	0	0	0	0	0	0	0	0	15	15
	Space	198.75	200	200	200	199.5	194	184	186.5		

Table B.1 Comparison of the bin-packing and the pattern-based models

	Bin-packing model	Pattern-based model
Unknown variables	$x_{ij} \in Z_+^1;$ $y_j \in \{0,1\}$	$x_j \in Z_+^1$
Bin j/Pattern j	$(x_{1j}, x_{2j}, \ldots, x_{mj})^T$	$(a_{1j}, a_{2j}, \ldots, a_{mj})^T$
Bin j packing/Pattern j width	$\sum_{i \in I} w_i x_{ij}$	$\sum_{i \in I} w_i a_{ij}$
Number of items i that must satisfy the ordered amount	$\sum_{j \in J} x_{ij}$	$\sum_{j \in J} a_{ij} x_j$
Number of bins/Number of sets	$\sum_{j \in J} y_j$	$\sum_{j \in J} x_j$

Proof of Theorem 2.2

The table below (Table B.1) states the equivalence between the two alternative formulations of the CSP.

It means every solution of the bin-packing model is a solution of the pattern-based model, and the reverse is also true. Q.E.D.

Proof of Theorem 2.3

We can prove the theorem by contradiction. Let us assume that the maximum is achieved in a point t^* not belonging to the above points. All inverse sawtooth functions have a negative slope in the small neighborhood of this point t^*. Thus, the sum also has a negative slope. So, shifting slightly to the left gives us a larger value of the sum function. It contradicts our initial assumption. Q.E.D.

Proof of Theorem 2.4

First, we notice that the optimal value of the objective function reaches every peak of the sum of the inverse sawtooth functions. Indeed, if it is not true, we can indicate a peak with the highest value of the summation function and a peak with the lowest value of the summation function (Fig. B.1).

Shifting the lowest component of the summation function slightly to the left, we reduce the highest peak and increase the lowest peak. It contradicts our assumption.

Second, we will prove that equidistant phase shifts deliver the optimal solution. Let us consider a pair of adjacent peaks at points $t = \varphi_k$ and $t = \varphi_{k+1}$ (Fig. B.2).

The inventory value for $t = \phi_k$:

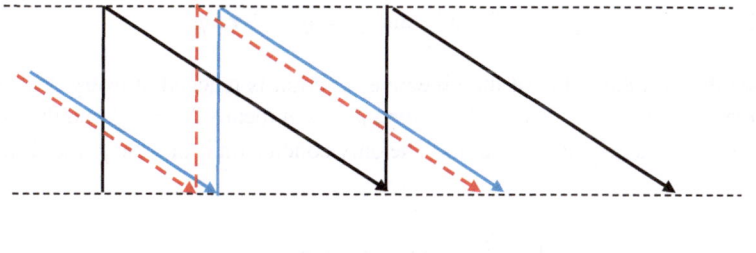

Fig. B.1 The highest summation peak comes from the black component, and the lowest comes from the blue component; the slight deviation of the blue component is shown in red

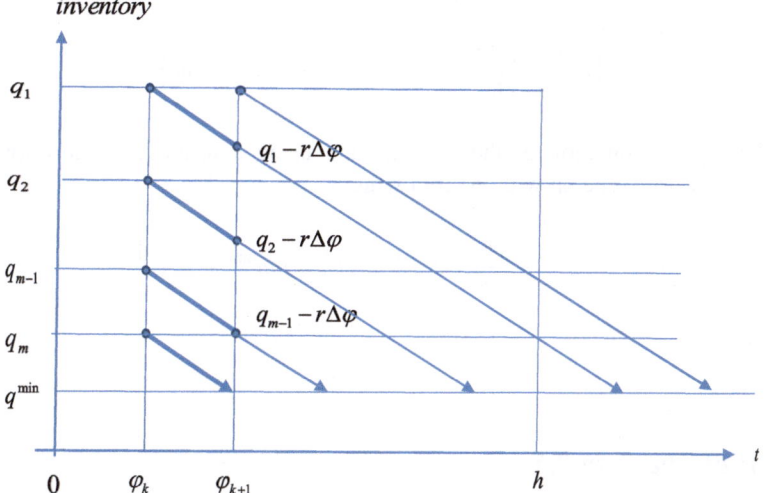

Fig. B.2 A pair of adjacent peaks

$$inventory_1 = \sum_{i=1}^{m} q_i$$

The inventory value for $t = \varphi_{k+1}$:

$$inventory_2 = q_1 + \sum_{i=1}^{m-1} (q_i - r\Delta\varphi)$$
$$= q_1 + \sum_{i=1}^{m} q_i - r\Delta\varphi(m-1) - q_m = inventory_1 + q^{min} + x^{max}$$
$$- r\Delta\varphi(m-1) - q^{min} - r\Delta\varphi = inventory_1 + x^{max} - mr\Delta\varphi.$$

Here, $\Delta\varphi = \varphi_2 - \varphi_1$, $r = x^{max}/h$, and we replaced

$$q_1 = q^{\min} + x^{\max} \text{ and } q_m = q^{\min} + r\Delta\varphi.$$

Since the optimal value of the objective function is reached at every peak, then $inventory_1 = inventory_2$. Thus, $x^{\max} - mr\Delta\varphi = 0$. It means $\Delta\varphi = \frac{h}{m}$. And this is true for every pair of adjacent time points. The only conclusion is that the phase shifts are equidistant:

$$\varphi_i^0 = \frac{h}{m}(i-1), \quad i = 1, 2, \ldots, m.$$

Third, according to our convention $\phi_1 = 0$. So, the peak value of the first item at the time $t = 0$ is $q^{\min} + x^{\max}$. Since the phase shifts are equidistant, the curves intersect the vertical line at every peak $t = \phi_i^0$ with values:

$$\left\{ q^{\min} + \frac{x^{\max}}{m}(m-k+1), \quad k = 1, \ldots, m \right\}.$$

Making a summation of these points and using a formula of the arithmetic progression, we come up with the following:

$$\sum_{k=1}^{m} \left(q^{\min} + \frac{x^{\max}}{m}(m-k+1) \right) = mq^{\min} + \frac{m+1}{2}x^{\max}.$$

Q.E.D.

Proof of Theorem 2.5

If the Least Common Multiple $T^{lcm} = LCM(T_1, T_2, \ldots, T_m)$ exist, then the sum of the inverse sawtooth functions is a periodic with a period T^{lcm} (Radcliffe 2013). So, the maximum of the sum will be achieved in the range $[0, T^{lcm})$. Q.E.D.

Proof of Theorem 2.6

Let φ^0 be an optimal solution to the problem $\mathbf{P} = (\mathbf{q}^{\min}, \mathbf{x}^{\max}, \mathbf{h}, \mathbf{w})$. Note that

$$\frac{t - \varphi_i}{h_i} = \frac{\lambda t - \lambda\varphi_i}{\lambda h_i}, \quad i \in I$$

With time adjustment, we can conclude that $\lambda\phi^0$ is the optimal solution to the problem $\mathbf{P}_\lambda = (\mathbf{q}^{\min}, \mathbf{x}^{\max}, \lambda\mathbf{h}, \mathbf{w})$, and the optimal value is the same. Q.E.D.

Appendix C: Discrete Warehouse Storage Space Model in Python 3

Figure C.1 presents Python code for the WSSP discrete model.

```
###############################################################################
# OptInvDisc.py
###############################################################################

import sys
import csv
from datetime import datetime
import copy
import math
from ortools.linear_solver import pywraplp
#from ortools.init import pywrapinit
import MyLibrary

#=============================================================================
# Class OptInvDisc.
#=============================================================================

class OptInvDisc:

        #-----------------------------------------------------------------------
        # OptInvDisc: Constructor
        #-----------------------------------------------------------------------

        def __init__(self,fileNameIn,fileNameConfig,fileNameOut = None):
                self.fileNameIn  = fileNameIn # (in) input data file name = set
                self.fileNameOut = fileNameOut       # (in) output file name
                self.fileNameConfig = fileNameConfig # (in) config file name
                self.orders = set()              # (in) set of orders
                self.qMin   = {}                 # (in) dict: order -> qMin[order]
                self.xMax   = {}                 # (in) dict: order -> xMax[order]
                self.h      = {}                 # (in) dict: order -> h[order]
                self.w      = {}                         # (in) dict: order -> w[order]
                self.tMax   = 0.0                        # (in) tMax
                self.dt     = 0.0                # (in) dt
                self.config = {}                 # (in) config
                self.isRead = False

        #-----------------------------------------------------------------------
        # OptInvDisc: Read the Config.csv file.
        #-----------------------------------------------------------------------

        def ReadConfig(self,traceFile):
                t = str(datetime.now())
                traceFile.write(t + " ReadConfig " + self.fileNameConfig + "\n")
                print(t + " ReadConfig " + self.fileNameConfig)

                try:
                        fileObject = open(self.fileNameConfig, 'r')
                except Exception as e:
                        print("Failed to open file: %s" % (str(e)))
                        return False

                csv_reader = csv.reader(fileObject, delimiter = ',', \
                quotechar = '"')
                rowCount = 0
                table = ""

                for row in csv_reader:
                        if(row):
                                if row[0] in {"Constraint","Criterion","Solver"}:
                                        table = row[0]
                                        if table == "Constraint":
                                                self.config[table] = set()
                                        elif table == "Criterion":
```

Fig. C.1 Python code for WSSP discrete model

```
                                      self.config[table] = {}
                          elif table == "Solver":
                                      self.config[table] = ""

                    elif (table == "Constraint") and (row[1] == "1"):
                          self.config[table].add(row[0])
                    elif (table == "Criterion") and (row[1] == "1"):
                          self.config[table][row[0]] = {}
                          self.config[table][row[0]]["Goal"] = row[2]
                          self.config[table][row[0]]["Rank"] = int(row[3])
                          self.config[table][row[0]]["Tolerance"] = float(row[4])
                    elif (table == "Solver") and (row[1] == "1"):
                          self.config[table] = row[0]

                    rowCount += 1

        fileObject.close()
        self.isRead = True
        print('...read ' + str(rowCount) + ' lines.')
        traceFile.write("...read " + str(rowCount) + " lines.\n")

        return True

#--------------------------------------------------------------------------
# OptInvDisc: Read the Input.csv file.
#--------------------------------------------------------------------------

def ReadInput(self,traceFile):
        t = str(datetime.now())
        traceFile.write(t + " ReadInput " + self.fileNameIn + "\n")
        print(t + " ReadInput " + self.fileNameIn)

        try:
                fileObject = open(self.fileNameIn, 'r')
        except Exception as e:
                print("Failed to open file: %s" % (str(e)))
                return False
        skipLine = True                           # skip the header of the csv file.

        csv_reader = csv.reader(fileObject, delimiter = ',', \
quotechar = '"')
        rowCount = 0

        for row in csv_reader:
                if( skipLine == False):
                        order           = row[0]
                        self.orders.add(order)
                        self.qMin[order]    = float(row[1])
                        self.xMax[order]    = float(row[2])
                        self.h[order]       = float(row[3])
                        self.w[order]       = float(row[4])
                        self.tMax           = float(row[5])
                        self.dt             = float(row[6])
                else:
                        skipLine = False
                rowCount += 1

        fileObject.close()
        self.isRead = True
        print('...read ' + str(rowCount) + ' lines.')
        traceFile.write("...read " + str(rowCount) + " lines.\n")
        return True
#--------------------------------------------------------------------------
# OptInvDisc: Class Solution.
#--------------------------------------------------------------------------

class Solution:
        def __init__(self,q,qInd,z,zInd,T,orders,w,dt):
                self.T      = T
                self.orders = orders
                self.w      = w
```

Fig. C.1 (continued)

```
                         self.dt        = dt

                         self.volume,self.wMax,self.wMin,self.phis = \
                         self.ComputeSolution(q,qInd,z,zInd)

              def ComputeSolution(self,q,qInd,z,zInd):
       # compute volume
                         volume = {}
                         for order in sorted(self.orders):
                                 volume[order] = {}
                                 for t in self.T:
                                          i = qInd[order][t]
                                          volume[order][t] = \
                    self.w[order] * q[i].solution_value()

       # compute criteria
                         wMax = 0.0
                         wMin = float("inf")
                         for t in self.T:
                                 s = 0.0
                                 for order in sorted(self.orders):
                                          i = qInd[order][t]
                                          s += volume[order][t]

                                 wMax = max(wMax,s)
                                 wMin = min(wMin,s)

       # compute phases
                         phis = {}
                         for order in sorted(self.orders):
                                 for t in self.T:
                                          i = zInd[order][t]
                                          if z[i].solution_value() == 1:
                                                  phis[order] = t * self.dt
                                                  break

                         return volume,wMax,wMin,phis

              def WriteSolution(self,fileObject):
       # criteria
                         fileObject.write("wMax," + str(self.wMax) + "\n")
                         fileObject.write("wMin," + str(self.wMin) + "\n")

       # title
                         fileObject.write("period,time")
                         for order in sorted(self.orders):
                                 fileObject.write("," + order)
                         fileObject.write("," + "total" + "\n")

       # volume
                         for t in self.T:
                                 fileObject.write(str(t))
                                 fileObject.write("," + str(t * self.dt))
                                 s = 0.0
                                 for order in sorted(self.orders):
                                          value = self.volume[order][t]
                                          fileObject.write("," + str(value))
                                          s += value
                                 fileObject.write("," + str(s))
                                 fileObject.write("\n")
                         fileObject.write("\n")

       # phases
                         fileObject.write(",phis")
                         for order in sorted(self.orders):
                                 fileObject.write("," + str(self.phis[order]))
                         fileObject.write("\n")
                         fileObject.write("\n")

#---------------------------------------------------------------------------
# OptInvCont: Model
```

Fig. C.1 (continued)

```
#---------------------------------------------------------------------------

def Model(self,traceFile ):
        traceFile.write("Running Model.\n")
    #-----------------------------------------------------------------------
    # Create the MIP solver.
    #-----------------------------------------------------------------------
        solverName = self.config["Solver"]
        print("Solver = ", solverName)
        solver = pywraplp.Solver.CreateSolver(solverName)

    #-----------------------------------------------------------------------
    # Calculate demand per time period dt.
    #-----------------------------------------------------------------------
        demand = {}
        for order in sorted(self.orders):
                demand[order] = self.xMax[order] * self.dt / self.h[order]

    #-----------------------------------------------------------------------
    # Create a list of time periods.
    #-----------------------------------------------------------------------
        numPeriods = int(math.floor(self.tMax/self.dt)) + 1
        hPeriods = {}
        for order in sorted(self.orders):
                hPeriods[order] = int(math.floor(self.h[order]/self.dt))

        T = list(range(numPeriods))
        T1 = [-1] + T

    #-----------------------------------------------------------------------
    # Create the variables.
    #-----------------------------------------------------------------------

    # Create x[i] ~ x[i,t] variables (arrivals)
        x = []
        xInd = {}
        i = 0
        for order in sorted(self.orders):
                xInd[order] = {}
                for t in T:
                        x.append(solver.NumVar(0,self.xMax[order],'x'))
                        xInd[order][t] = i
                        i += 1

    # Create y[i] ~ y[i,t] variables (departure)
        y = []
        yInd = {}
        i = 0
        for order in sorted(self.orders):
                yInd[order] = {}
                for t in T:
                        y.append(solver.NumVar(0,float('inf'),'y'))
                        yInd[order][t] = i
                        i += 1

    # Create q[i] ~ q[i,t] variables (inventory)
        q = []
        qInd = {}
        i = 0
        for order in sorted(self.orders):
                qInd[order] = {}
                for t in T1:
                        q.append(solver.NumVar(0,float("inf"),'q'))
                        qInd[order][t] = i
                        i += 1

    # Create z[i] ~ z[i,t] binary variables (indication of arrivals)
        z = []
        zInd = {}
        i = 0
        for order in sorted(self.orders):
```

Fig. C.1 (continued)

```
                        zInd[order] = {}
                        for t in T:
                                z.append(solver.IntVar(0, 1, 'z'))
                                zInd[order][t] = i
                                i += 1

# Create continious variables - max volume and min volume
        wMax = solver.NumVar(0,float('inf'),'wMax')
        wMin = solver.NumVar(0,float('inf'),'wMin')

#------------------------------------------------------------------------
# Create constraints
#------------------------------------------------------------------------

        cstr = []

        for constraint in self.config["Constraint"]:

#------------------------------------------------------------------------
# q[i,t] = q[i,t-1] + x[i,t] - y[i,t] for all i and t.
#------------------------------------------------------------------------
                if constraint == "InventoryDynamics":
                        for order in sorted(self.orders):
                                for t in T:
                                        i = qInd[order][t]
                                        j = qInd[order][t-1]
                                        k = xInd[order][t]
                                        l = yInd[order][t]
                                        cstr.append(solver.\
                    Constraint(0,0,'InventoryDynamics'))
                                        cstr[-1].SetCoefficient(q[i], 1)
                                        cstr[-1].SetCoefficient(q[j], -1)
                                        cstr[-1].SetCoefficient(x[k], -1)
                                        cstr[-1].SetCoefficient(y[l], 1)

#------------------------------------------------------------------------
# x[i,t] = xMax[i] * z[i,t], for all i and t.
#------------------------------------------------------------------------
                elif constraint == "QtyProductArrival":
                        for order in sorted(self.orders):
                                for t in T:
                                        cstr.append(solver.\
                    Constraint(0,0,'QtyProductArrival'))
                                        i = xInd[order][t]
                                        j = zInd[order][t]
                                        cstr[-1].SetCoefficient(x[i],1)
                                        cstr[-1].SetCoefficient(z[j],-self.xMax[order])

#------------------------------------------------------------------------
# y[i,t] = demand[i], for all i and t.
#------------------------------------------------------------------------
                elif constraint == "QtyProductDeparture":
                        for order in sorted(self.orders):
                                for t in T:
                                        cstr.append(solver.\
                    Constraint(demand[order],\
                    demand[order],'QtyProductDeparture'))
                                        i = yInd[order][t]
                                        cstr[-1].SetCoefficient(y[i],1)

#------------------------------------------------------------------------
# sum{i,tau=t,t+h[i]|z[i,tau]} <= 1, for all i and t=0,numPeriods-h[i]
#------------------------------------------------------------------------
                elif constraint == "OneArrivalPerCycle":
                        for order in sorted(self.orders):
                                for t in T:
                                        if t < numPeriods - hPeriods[order]:
                                                cstr.append(solver.Constraint(0,1,\
                    'OneArrivalPerCycle'))

                                                for k in range(hPeriods[order]):
                                                        tau = t + k
```

Fig. C.1 (continued)

```
                                            i = zInd[order][tau]
                                            cstr[-1].SetCoefficient(z[i],1)

#----------------------------------------------------------------------
# wMax >= sum{ i in I | w[i] * q[i,t]}  for all t.
#----------------------------------------------------------------------
             elif constraint == "MaxInventoryVolume":
                    for t in T:
                            cstr.append(solver.Constraint(0,float('inf'),\
            'MaxInventoryVolume'))
                            cstr[-1].SetCoefficient(wMax, 1)
                            for order in sorted(self.orders):
                                    i = qInd[order][t]
                                    cstr[-1].SetCoefficient(q[i],-self.w[order])

#----------------------------------------------------------------------
# wMin <= sum{ i in I | w[i] * q[i,t]}  for all t.
#----------------------------------------------------------------------
             elif constraint == "MinInventoryVolume":
                    for t in T:
                            cstr.append(solver.Constraint(-float('inf'),0,\
            'MinInventoryVolume'))
                            cstr[-1].SetCoefficient(wMin, 1)
                            for order in sorted(self.orders):
                                    i = qInd[order][t]
                                    cstr[-1].SetCoefficient(q[i],-self.w[order])

#----------------------------------------------------------------------
# q[i,t] >= qMin[i] + demand[i] for all i and t.
#----------------------------------------------------------------------
             elif constraint == "InventoryLowerBound":
                    for order in sorted(self.orders):
                            rhs = self.qMin[order] + demand[order]
                            for t in T:
                                    cstr.append(solver.Constraint(rhs, float("inf"),\
            'InventoryLowerBound'))
                                    i = qInd[order][t]
                                    cstr[-1].SetCoefficient(q[i], 1)

#----------------------------------------------------------------------
# q[i,t] <= qMin[i] + xMax[order] for all i and t.
#----------------------------------------------------------------------
             elif constraint == "InventoryUpperBound":
                    for order in sorted(self.orders):
                            rhs = self.qMin[order] + self.xMax[order]
                            for t in T:
                                    cstr.append(solver.Constraint(0,rhs,\
            'InventoryUpperBound'))
                                    i = qInd[order][t]
                                    cstr[-1].SetCoefficient(q[i], 1)

#----------------------------------------------------------------------
# sum{ i in I | z[i,t] <=1 }  for all t.
#----------------------------------------------------------------------
             elif constraint == "ArrivalTimeSpread":
                    for t in T:
                            cstr.append(solver.\
            Constraint(0,1,'ArrivalTimeSpread'))
                            for order in sorted(self.orders):
                                    i = zInd[order][t]
                                    cstr[-1].SetCoefficient(z[i], 1)

#----------------------------------------------------------------------
# Solve the model
#----------------------------------------------------------------------

# create a working dictionary: "crit -> rank", and sort it by ranks.
      critRank = {}
      for criterion in self.config["Criterion"]:
            critRank[criterion] = \
            self.config["Criterion"][criterion]["Rank"]
```

Fig. C.1 (continued)

```
                    critRank = MyLibrary.SortDictByValue(critRank)

          solutions = []

     # lexicographic optimization
          passCrit = 0
          for criterion in critRank:
                    print("criterion=", criterion)
     # create an objective function
                    objective = solver.Objective()
                    if self.config["Criterion"][criterion]["Goal"] == "min":
                              objective.SetMinimization()
                    elif self.config["Criterion"][criterion]["Goal"] == "max":
                              objective.SetMaximization()
                    if criterion == "MinMaxVolume":
                              objective.SetCoefficient(wMax, 1)
                              objective.SetCoefficient(wMin, 0)
                    elif criterion == "MaxMinVolume":
                              objective.SetCoefficient(wMin, 1)
                              objective.SetCoefficient(wMax, 0)

                    print('Num of variables =', solver.NumVariables())
                    print('Num of constraints =', solver.NumConstraints())
                    t = str(datetime.now())
                    traceFile.write(t + " Start solving the model..." + "\n")
                    print(t + " Start solving the model...")

                    status = solver.Solve()

                    if status == pywraplp.Solver.OPTIMAL:
                              print('Solution:')
                              print('Objective value =', objective.Value())

          # accumulate solution
                              solutions.append(OptInvDisc.\
          Solution(q,qInd,z,zInd,T,self.orders,self.w,self.dt))

          # add upper/lower bound for min/max criterion
                              if passCrit < len(critRank) - 1:
                                        tol = self.config["Criterion"][criterion]["Tolerance"]
                                        if criterion == "MinMaxVolume":
                                                  cstr.append(solver.Constraint(0,objective.Value() +\
          tol,'MinMaxVolume'))
                                                  cstr[-1].SetCoefficient(wMax, 1)
                                        elif criterion == "MaxMinVolume":
                                                  cstr.append(solver.Constraint(objective.Value() -\
          tol,float("inf"), 'MaxMinVolume'))
                                                  cstr[-1].SetCoefficient(wMin, 1)

                    else:
                              print('The problem does not have an optimal solution.')

                    passCrit += 1

#-----------------------------------------------------------------------
# Output solution
#-----------------------------------------------------------------------

     print('Elapsed time = %f milliseconds' % solver.wall_time())
     print('Number of iterations = %d' % solver.iterations())
     print('Branch-and-bound nodes = %d' % solver.nodes())
     return solutions

#-----------------------------------------------------------------------
# OptInvDisc: WriteSolutions
#    Input: self,solutions
#-----------------------------------------------------------------------

def WriteSolutions(self,traceFile,solutions):

     t = str(datetime.now())
```

Fig. C.1 (continued)

```
                    traceFile.write(t + " WriteSolutions " + self.fileNameOut + "\n")
                    print(t + " WriteSolutions " + self.fileNameOut)

                    try:
                            fileObject = open(self.fileNameOut, 'w')
                    except Exception as e:
                            print("Failed to open file: %s" % (str(e)))
                            return False

                    for solution in solutions:
                            solution.WriteSolution(fileObject)

                    fileObject.write("\n")
                    fileObject.close()

        #-------------------------------------------------------------------------
        # OptInvCont: RunProgram
        #-------------------------------------------------------------------------

        def RunProgram(self,traceFile,alg):
                self.ReadInput(traceFile)
                self.ReadConfig(traceFile)
                solutions = self.Model(traceFile)
                self.WriteSolutions(traceFile,solutions)

        #-------------------------------------------------------------------------
        # OptInvCont: Destructor
        #-------------------------------------------------------------------------

        def __del__(self):
                print("Destructor called, an OptInvDisc is deleted.")

#################################################################################
# End of the code
#################################################################################
```

Fig. C.1 (continued)

References

AIMMS Optimization modeling (2023). https://documentation.aimms.com/_downloads/AIMMS_modeling.pdf

AMPL Development (2022–2023). https://dev.ampl.com

Bazaraa MS, Jarvis JJ, Sherali HD (1990) Linear programming and network flows. Wiley, New York

Bellman R (1957) Dynamic programming. Princeton University Press, Princeton, NJ

Benders JF (1962) Partitioning procedures for solving mixed-variables programming problems. Numer Math 4(3):238–252

Ben-Tal A, Nemirovski A (1999) Robust solutions of uncertain linear programs. Oper Res Lett 25(1):1–13

Carlson B et al (2012) MISO unlocks billions in savings through the application of operations research for energy and ancillary service markets. Interfaces 42(1):58–73

Chvatal V (1983) Linear programming. W.H. Freeman and Company, New York

Conejo AJ, Castillo E, Mínguez R, García-Bertrand R (2006) Decomposition techniques in mathematical programming. Springer, New York

De Carvalho JMV, Rodrigues AJG (1995) An LP-based approach to a two-stage cutting stock problem. Eur J Oper Res 84(5):580–589

FICO Xpress Mosel (2017) Users' guide, Release 4.8, 2017. https://www.fico.com/fico-xpress-optimization/docs/latest/mosel/UG/dhtml/GUID-5DBA5E00-B9BC-3A85-A9AF-A03B2F564C43.html

FICO Xpress Optimizer Reference manual. https://www.fico.com/fico-xpress-optimization/docs/latest/solver/optimizer/HTML/GUID-3BEAAE64-B07F-302C-B880-A11C2C4AF4F6.html

Fisher ML (1985) An applications-oriented guide to Lagrangian relaxation. Interfaces 15(2):10–21

Gilmore PC, Gomory RE (1961) A linear programming approach to the cutting stock problem. Oper Res 8:849–859

Gilmore PC, Gomory RE (1963) A linear programming approach to the cutting stock problem. Part 2, vol 11. Oper Res, pp 838–888

Google OR-Tools Route. Schedule. Plan. Assign. Pack. Solve. https://developers.google.com/optimization

Gurobi Optimizer Reference Manual (2020). https://www.gurobi.com/documentation/current/refman/index.html

Haessler RW (1968) An application of heuristic programming to a nonlinear cutting stock problem occurring in the paper industry. The University of Michigan, Ph.D. doctoral dissertation

Hedman KW, O'Neil RP, Oren SS (2009) Analyzing valid inequalities of the generation unit commitment problem. IEEE. http://www.ieor.berkeley.edu/~oren/pubs/Analyzing2009.pdf. Retrieved 5 October 2009

IBM ILOG CPLEX Optimization Studio (2017). https://www.ibm.com/products/ilog-cplex-optimization-studio. https://www.ibm.com/docs/en/icos/22.1.0?topic=apis-concert-technology-c-users

Johnson MP, Rennick C, Zak E (1997) Skiving addition to the cutting stock problem in the paper industry. SIAM Rev 39(3):472–483

Johnson MP, Rennick C, Zak E (1999) One-dimensional cutting stock problem in just-in-time environment. Pesqui Oper 19(2):145–158

Kantorovich LV (1939) Mathematical methods for production planning and optimization. Leningrad University, Leningrad. (in Russian). [English translation: Management Science, 6, 1960, pp. 366–422]

Kellerer H, Pferschy U, Pisinger D (2004) Knapsack problems. Springer, Berlin

Lavigne JR (1993) Pulp & paper dictionary. Miller Freeman Books, San Francisco

MAJIQTRIM: Trim optimization software (1995) Users' manual. Majiq Systems and Software, Redmond, WA

Marcotte O (1985) The cutting stock problem and integer rounding. Math Program 33:82–92

Martinovich J (2022) A note on the integrality gap of cutting and skiving stock instances. Q J Oper Res 20:85–104

Nazareth JL (2004) An optimization primer. Springer, New York

Optimization Online (2022) The SCIP optimization suite 8.0. https://optimization-online.org/2021/12/8728/

Radcliffe D (2013) Sum of periodic functions. https://mathblag.wordpress.com/2013/09/01/sums-of-periodic-functions/

Rajan D, Takriti S (2005) Minimum up/down polytopes of the unit commitment problem with start-up costs. IBM research report

Ruiz PA, Philbrick CR, Zak E, Cheung KW, Sauer PW (2009) Uncertainty management in the unit commitment problem. IEEE Trans Power Syst 24(2):642–651

Scheithauer G, Terno J (1995) The modified integer round-up property of the one-dimensional cutting stock problem. Eur J Oper Res 3(84):562–571

Shapiro A, Dentcheva D, Ruszczynski A (2009) Lectures on stochastic programming, modeling and theory. SIAM, Philadelphia

StackExchange Network (2019) Modeling floor function exactly. https://or.stackexchange.com/questions/443/modeling-floor-function-exactly/505#505

Williams HP (2013) Model building in mathematical programming, 5th edn. Wiley, New York

Wolsey LA (1998) Integer programming. Wiley, New York

Zak E (2002a) Modeling multistage cutting stock problem. Eur J Oper Res 141(2):313–327

Zak E (2002b) Row and column generation technique for a multistage cutting stock problem. Comput Oper Res 29:1143–1156

Zak E (2003) The skiving stock problem as a counterpart of the cutting stock problem. Int Trans Oper Res 10(6):637–650

Zak E (2006) Is integrality gap in cutting stock problem less than 2? Discret Appl Math 154(3): 606–607

Zak E (2011) Lessons learned en route to the Edelman. ORMS Tod 38(3):36–38

Zak E (2018) Minimization of sum of inverse sawtooth functions. In: 23rd International Symposium on Mathematical Programming, Proceedings, p 141. https://ismp2018.sciencesconf.org/data/bookFullProgram.pdf

Zak E, Dereksdottir E (2001) A modified lexicographic algorithm for multi-constraint knapsack problems, Technical report OR-003/0644. Majiq Systems and Software, Redmond, WA

Zhao M, Kalagnanam J (2009) Branch-and-cut for the unit commitment problem. In: Proceedings, the sixth workshop on MIP, Berkley, p 14. chrome-extension:// efaidnbmnnnibpcajpcglclefindmkaj/https://atamturk.ieor.berkeley.edu/mip2009/mip200 9program.pdf.